ALLOBERDI M. MADRAKHIMOV

Development of technology for obtaining filler

ALLOBERDI M. MADRAKHIMOV

Development of technology for obtaining filler

ScienciaScripts

Imprint

Any brand names and product names mentioned in this book are subject to trademark, brand or patent protection and are trademarks or registered trademarks of their respective holders. The use of brand names, product names, common names, trade names, product descriptions etc. even without a particular marking in this work is in no way to be construed to mean that such names may be regarded as unrestricted in respect of trademark and brand protection legislation and could thus be used by anyone.

Cover image: www.ingimage.com

This book is a translation from the original published under ISBN 978-620-6-17312-0.

Publisher:
Sciencia Scripts
is a trademark of
Dodo Books Indian Ocean Ltd. and OmniScriptum S.R.L publishing group

120 High Road, East Finchley, London, N2 9ED, United Kingdom
Str. Armeneasca 28/1, office 1, Chisinau MD-2012, Republic of Moldova, Europe
Printed at: see last page
ISBN: 978-620-7-72549-6

Contents

In this monograph is the development of non-waste technology for obtaining conditioned wood-fibre filler from cotton stalks and composite wood-plastic board materials from them.

The scientific novelty of the research consists in the following: the technological process of cotton stalk crushing for the purpose of obtaining wood fibre fillers for composite board materials has been developed, wood fibre fillers for the production of wood-plastic board materials have been obtained by carrying out the crushing process in two stages, the efficiency of the process and the high content of porous wire in the condensing mass and the relationship of porous rod to the transmission and cutting speed have been determined, a change in the design of the composite board materials has been made.

AuthorM. A.Madrakhimov

CURRENT STATE OF THE ART IN GRINDING AND PRODUCTION OF FILLERS FROM STEMS OF ANNUAL PLANTS AND THEIR APPLICATION IN THE PRODUCTION OF COMPOSITE WOOD-PLASTIC BOARD MATERIALS

§ 1.1 Status and analysis of annual plant stems and the potential use of cotton stems in the production of composite wood-plastic and other materials.

The production of wood-plastic materials in the world is developing intensively from year to year. Valuable properties of wood-plastic board materials, such as homogeneity of microstructure and properties in different directions in volume and plane, relatively small changes in dimensions under conditions of parchment moisture gives a wide opportunity for their production. Comparatively easy processability of obtaining products of various configurations, shape of parts and sheet materials of large formats, as well as the possibility of using available polymer binders for them favours a wider use of annual plant stems for the production of the necessary materials [2; p.28-29].

The ever-increasing volume of construction already now consumes about half of the total volume of wood consumed. In particular, more than 300 thousand m^3 of composite wood-plastic board materials are consumed annually in our country. Of these, almost 250 thousand m^3 are imported from abroad for foreign currency [2; p.35].

Due to the limited forest resources, both in Uzbekistan and in other countries of Central Asia, a tendency to use agricultural waste and stems of various annual plants as raw materials for the manufacture of wood-plastic board materials has emerged: stems of flax and hemp bark, husks, stems of cotton, rice, sunflower husks, coffee husks, ground nuts, coconut palms, bamboo stems [3; p. 89-93]. It is known that their reserves are quite large and are annually renewed.

Production of wood-plastic board materials from the stems of annual plants in the country has not yet found its wide development. However, in a number of foreign countries barks of bast plants - flax, hemp and kenaf - are widely used as raw materials for filling composite wood-plastic board materials. At present, boards made of bark are produced in Poland, in France and in a number of other countries. Thus, out of 60 grades of boards produced in France - 12 grades of boards made of bark. Boards from bagasse - sugar cane waste, are produced in Cuba, Argentina, Philippines, India, Brazil, Mexico, South Africa and other countries [4; p. 180-210].

Composite wood-plastic boards have rather high physical and mechanical properties: specific weight - 700-800 kg/m^3 , bending strength - 32,0-33,0 MPa, transverse tensile strength -71,0 MPa, coefficient of linear expansion - 0,3%, lateral resistance to nail pulling 90153 N/mm [5; p. 10-21, 102-120].

There is some experience in obtaining composite wood-plastic board materials from reed stalks. Bamboo stalks, vine trimmings and various types of straw are also used as filler material replacing wood [6; p. 112-164].

It can be seen from a number of works that linseed bark is mainly used as a filler to produce composite wood-plastic board materials. Bark from the flax plant is obtained sufficiently crushed and dry, which simplifies the production of wood-plastic boards and is an advantage compared to wood [7; p. 282-311].

The particle size of flaxseed bark (length 1-3 cm, width 2-3 mm and thickness 0.10.3 mm) allows to obtain composite wood-plastic boards with a smooth surface. The chemical composition of flaxseed bark is presented in Table 1.1 [8; p. 5-14].

Table 1.1.

Chemical composition of flaxseed meal

Components	Content, %
Cellulose	3,3
Pentosans	33
Lignin	21
Ash	1,2
Water-soluble substances	5,83
Substances soluble in ether	1,22
Moisture	up to 8

The tensile strength is 22.0 MPa.

There is some experience in obtaining composite wood-plastic boards from reed (reed) stalks. It is known that a reed stem consists of an outer cylinder and a central air-bearing cavity. The cylinder wall is covered with a layer of films covered with fatty substance, inside the wall the stalks are covered with wax-like plaque and therefore reed stalks have the property of significant water absorption. Composite wood-plastic boards made with the addition of hydrophobising emulsion have a water absorption index of about 3050 %. Cane stalks contain: cellulose 47-49 %, lignin 22-24 %. Due to high water absorption, boards made of reed stalks have not been widely used yet [9;p.5O].

In [10; pp. 1-3,   11;   pp. 49-59], the results of the following were reported research on the use of bagasse (sugarcane waste) in paper production. These studies have found that bagasse combined with wood chips can also be used to produce composite wood-plastic materials and boards. Bagasse is currently being processed in Argentina, the Philippines, India, Brazil, Taiwan, Mexico and the Dominican Republic. Of interest is the experience of Cuba in using bagasse for this purpose. Fibres, the content of which is 50-80%, are the main component providing high strength properties of composite wood-plastic board materials made of bagasse [13; p. 42]. A distinctive feature of the technological process of manufacturing wood-plastic board materials from bagasse is the organisation of storage of raw materials, the need to separate small particles from large ones, as well as certain difficulties arising in the dosing of particles, moulding and underpressing of the mat. Bagasse usually has a volumetric weight of 160 kg/m^3 , a moisture content of 50 % and contains sucrose.

Dry bagasse is 64 % fibre, 24 % heartwood and 10 % impurities. Strength characteristics of plates with density 530-620 kg/m^3 have the following indices - bending strength - 18,0 MPa, perpendicular plate tension - 3,5 kgf/cm^2 [14; p. 78-79].

It is known [15; p. 22] that in order to obtain sufficiently strong thin composite wood-plastic materials and boards from bamboo stalks, the particles of its grinding should have a flat shape and such optimal dimensions: length - 20-30 mm, width - 10-12 mm and thickness - 0.2-0.4 mm, as well as the initial moisture content of the particles of about 4-5%   . Physical and mechanical properties

composite wood-plastic board materials using chopped bamboo are given in Table 1.2 [16; 182-200].

Boards based on shredded bamboo appear to be sufficiently strong in the case of thin sheet material (3-6 mm thickness) and in many cases can replace glued plywood.

Table 1.2.

**Physical and mechanical properties of composite
wood-plastic board materials based on bamboo filler**

Name of indicators	Limit of values of the Indicator
Volume weight, y, g/cm^3	0,73-0,77
Water absorption in 24 hours, w, %	45-50
Swelling in thickness, AS, %	10-14
Static bending strength, about$_и$, MPa	43,2-43,3
Tensile strength, about$_p$, MPa	22,0-25,5

A number of works [20; Patent] are devoted to obtaining board materials from grapevine prunings. The possibility of obtaining such boards has been proved and physical and mechanical parameters for boards made of both pure grapevine and with the addition (up to 50 %) of wood shavings are given. Tables 1.3 contain data on the physical and mechanical properties of composite wood-plastic board materials made of grapevine [21; p. 163, 22; p. 87-88].

Table 1.3.

**Physical and mechanical properties of composite wood fibre composites
of plastic slab materials made of grapevine**

Indicators	Grapevine slabs	Grapes, vine + 20% ancient, shavings.	Grapes, vine + 50% ancient, shavings.
Volume weight, y, kg/cm^3	750	760	741
Ultimate strength at static bending, about$_и$, MPa	9,5-16,5	23,8	17,6
Ultimate strength at stretching perpendicular to the plate, about$_p$, MPa	0,5-0,72	0,38	0,50
Water absorption, w, %	67-73	-	-
Swelling, AS, %	17-25	10,1	19,1

Note that there are also a number of ways to obtain composite wood-plastic board materials from rice straw and other stems of annual plants [23; p. 97-98]. The properties of the boards are as follows: density 1.11 g/cm^3 , flexural strength 19.2 MPa, water absorption 27.9 %, swelling 21.3 % [24; p. 100-162].

Of the materials listed above, cotton stalks are the closest to wood, especially in structure and properties.

Further we will consider the possibility of cotton stalks application in wood-plastic boards and other materials. It is well known that cotton is one of the main well-studied technical plants [25; pp. 108-109, 26; pp. 102-108, 27; pp. 80-84, 28; pp. 2-5]. In our country, cotton crops occupy more than 40 % of the total area devoted to spinning plants. Cotton is cultivated mainly to produce fibre and fat: one ton of raw cotton produces, on average, 300 m of fabric and 100-110 kg of highly valuable edible oil [29; p. 127-131]. The main feature of cotton is that all parts of the plant find practical use [30; pp. 11-30, 107-118, 31; p. 978, 32; pp. 54-60]. Thus, down-lint, delint and others are used to obtain cellulose for the production of photographic film, varnishes, paper and ebonite. The husks and cake left after processing of cotton seeds for oil. And wastes of cotton cleaning industry are valuable fodder for livestock [33; pp. 5-12, 34; pp. 19-20, 35; p. 5].

The husks are widely used in the hydrolysis industry. From a tonne of husks it is possible to obtain 150 kg of furfural and a significant amount of ethyl alcohol. It is known that cotton leaves are not inferior to lemons in citric acid content (6.5-8.5 %) [36; p. 32-38, 34; p. 19]. Cotton stalks are also a valuable raw material [35; p. 22, 36; p. 14-15]. 12

According to experts' estimates, with the production level of raw cotton in Uzbekistan in the amount of 3.0 million tonnes, the so-called "waste" of cotton is at least 4.0-4.5 million tonnes. Over 70% of the waste is cotton stalks, which is more than 2.5-3.0 million tonnes. [37; p. 112-124], which is a very important raw material resource.

As early as 1909, at the Kamensk paper factory, an experimental batch of paper was produced from cotton stalks [38; p. 112-142].

At present cotton stalks are used as fuel for dwellings, and abroad also for electricity, steam [39; pp. 9-11,], as fertiliser [41; pp. 4-12], as raw material for hydrolysis plants [42; pp. 17, 43; pp. 287293, 44; pp. 6770, 107-109; pp. 183, 45; pp. 18-20], for paper production [46; pp. 3-8]. From a tonne of stalks, 90-120 kg of furfural, which has stimulating properties of compound fertiliser, can be obtained by hydrolysis. However, the main part of cotton stalks finds no application and is used by the population as fuel.

It should be pointed out that in our country the use of cotton stalks for the production of composite wood-plastic board materials for the furniture industry and construction, as well as for the production of parts for mechanical engineering is considered more appropriate [47; p. 10-14].

Uzbekistan is a rich country in cotton production, so we have a lot of cotton stalks.

And from 1.8 tonnes of stalks 1 m^3 of building board is obtained, which is equal to 2 m^3 of sawn timber or 4 m^3 , round wood [48; p. 150-152]. Calculations show that using about 40-50 % of cotton stalks for the production of composite wood-plastic boards the need of the republic in them will be satisfied. However, for this purpose it is necessary to solve three groups of questions: firstly, harvesting of cotton stalks; secondly, production of wood-fibre fillers from cotton stalks and thirdly, production technology of composite wood-plastic board materials [49; p. 12-14]. The present dissertation work is devoted to the solution of the last two groups of questions [50; c. 126].

**§1.2 Study and analysis of the status of existing methods and
devices for grinding wood, wood-like and stems of
annual plants in order to obtain wood-fibre
fillers for composite wood-plastic board
materials from them**

Various methods and devices for grinding wood to obtain chips and shavings for the production of chipboard and wood-plastic board materials are known, which are described in detail in [51; p. 238-241].

Shredded wood particles as the main component of the board have a significant impact on the quality of wood-plastic board materials, which is also related to the economics of production. The cost of raw materials and the costs of chip production account for up to 50 % of the production cost [52; p. 6-8, 53; p. 22].

Various wood materials, from logs to sanding dust, are commonly used as raw materials for the production of composite boards. By processing wood or wood waste, a large number of wood particles with different geometric shapes and sizes can be obtained, suitable both for their direct use in the technological process of board production and for other purposes. The type of wood particles and the cost of obtaining them is an important factor in determining the economics of production. Each particle type, its inherent geometry and the combination of different particles have a determining influence on board quality [54; p. 101, 55; p. 18].

There are high requirements to the technology of raw material grinding, as the quality of the ground mass, particle size and shape, roughness 14 of the chip surface and their interrelation with each other are the defining characteristics of the board. The configuration and particle sizes have a dominant influence on such important properties of boards as bending strength, stiffness, tensile strength perpendicular to the layer, holding power of screws and nails, attitude to moisture, machinability on machine tools, etc. [56; p. 8-11,etc.]. [56; pp. 8-11,57; pp. 9-10].

It is well known that the size and shape of the pulverised wood particles have a direct impact on the operation of processing equipment. This includes the particle separation process and applies to conveyor equipment, dryers, mixers, moulding machines and even presses.

The geometry, fractional composition and quality of the obtained particles, the cost of their production mainly depend on the grinding method and on the machines used.

7

The following shredding methods exist for shredding annual plant stems, in particular cotton stems, in order to obtain fillers for the production of wood-plastic composite board materials: crushing, cutting, milling. All feed grinders and wood shredding machines for chipboard production are mainly based on these methods. It should be noted that existing machines [58] for shredding annual crop stems are mainly designed for fodder [59].

Currently, agriculture is used for the above purposes:

1. Rough fodder chopper IGK-ZOB;
2. Moskvichka hammer grain crusher (KDM2);
3. Universal fodder crusher "Ukraine" KDU-2,0;
4. Shredder IRT-165, etc.

These machines usually use the methods of grinding such as impact crushing, splitting, abrasion, flaking and cutting. The authors of a number of works have attempted to grind grapevine, rice straw, flax bark on these machines in order to obtain plates from them. There is a positive result. However, these machines are not suitable for cotton stalks. As it has already been noted, cotton stalks differ from stalks of other annual plants by the presence of a particularly strong and elastic bark, which is practically not crushed by the working bodies of feed crushers. Machines used for shredding wood are also not suitable for cotton stalks. Thus, machines based on raw material milling grind only the wood and core part of the stalk. The particles can have a very wide range of sizes and shapes. The bark of stalks under blows turns into a crushed fibrous mass, fibres in it can have a considerable length, clump and complicate secondary crushing, clog machine units, clog machine organs at subsequent stages of technology. Machines having knives and other cutting parts as a working organ are also not capable of shredding cotton stalks, as during the operation of these machines the bark is not cut, but mainly scraped off the stalk, also forming long fibre. In addition, the knife blades become enveloped and require frequent cleaning [60; patent, 61; pp. 108-109]. This work also indicates that cotton stalks were pre-crushed on a Volgar-5 forage crusher, with the addition of   10% wood chips, and   then were pulverised on a DS-3 chip grinding machine. However, the author does not explain how wood chips, with their relatively small content, affect the chopping process. The author also suggests freezing the stems before chopping. But this requires additional energy inputs and is feasible only when the stems are highly moist.

We made an attempt to modernise the existing serial equipment and use it for shredding cotton stalks for the production of particleboard, which is close to wood in quality [62; p. 27]. For this purpose, the following equipment was used: DS-2 chipboard machine Schwabedissen chipboard machine

chipping machine DS-3

universal crusher DU-2

veneer crusher for veneer of veneer vapour DSH.

6 variants of grinding were performed. 1 - variant - shredding on chip machine DS-2, designed to produce thin face wood chips. For this purpose, cotton stalks were first cut

into 30 cm long sections, stacked and pressed from above against the knife disc of the machine. Due to the lack of pile density and the presence of voids between the stalks, the productivity of the machine was much lower than for chopping wood. The presence of earth on the roots of the bushes and the presence of bark caused the knives to blunt quickly and the machine to become very hot. The resulting particles were 1-10 mm (60.0 %) and less than 1 mm (23 %), which does not meet the production requirements.

2 - option - shredding of cotton stalks on the Schwabedissen machine. The drum shredding machine of the German company "Schwabedissen" has as a cutting head a set of milling cutters with spirally arranged blades along the cylinder formation. On the knife shaft is located 6 rows of 28 blades in each. The blades cut chips 20 mm wide and 0.2 mm thick. The shredding of cotton stalks on the above machine also did not give positive results. The stalks were mostly threshed with fibres of too great length - 50-100 mm. Half of the pulverised mass was not suitable for consumption.

3 - variant - shredding of cotton stalks on a Schwabedissen machine with subsequent shredding on a chip machine DS-3, which is usually designed for processing chips produced in chippers from sawmill waste and thin-sized roundwood into chips. There are 26 knives located on the inner surface of the cup-shaped blade head shell of this 600 mm diameter machine. This machine processes chips up to 90 cm long and up to ZO mm wide into chips with a thickness of 0.2-1 mm.

When shredded on the Schwabedissen machine, the cotton stalks are transformed into fibres, and when subsequently shredded on the DS-3 machine - into dust-like particles from 1 to 10 mm.

4 - variant - chopping on a DS-3 chipping machine. For this purpose, the stems were first manually cut into 6-8 cm sections and then fed into the DS-3 machine. Sieve analysis showed that the particles were of the right thickness and length, the number of particles suitable for board production, i.e. greater than 1 mm in length, was more than 75 per cent.

5 - option - shredding in the DU-2 crusher and then on the DS-2 machine. The DU-2 machine is designed for processing of logging waste (branches, twigs, tops), sawmill waste (hump, lath, wood pieces up to 18 cm in diameter) into technological chips. In chipboard shops the DU-2 machine is also used for shredding veneer. For shredding cotton stalks machine DU-2 proved to be of little use, as it not only cut the stalks, but also a significant part of them is crushed into fibres. The resulting mass was fed to the DS-3 machine, where it was turned into shavings. Sieve analysis of the shavings showed that the required fraction was 85 %, but it contained 43 % of fine particles from 1 to 2 mm.

The 6th variant - shredding on the DSh crusher and on the DS-3 machine. Since the DU-2 crusher turns part of the stems into fibres, the DSh crusher was tested. After crushing in the crusher, particles over 10 mm make up 19 % of the total amount and therefore are subject to secondary crushing in the crusher. As this involves additional costs, this option was also found unsuitable for pulverisation.

It was found that the most acceptable is the shredding of cotton stalks on a DU-2 crusher with subsequent additional shredding on a DS-3 machine. However, it should be noted that even these machines cannot process a large amount of cotton stalks, as cotton bark and soil left on the roots of the stalks put out of order the cutting organs of the machines. In addition, the mass obtained is not of high quality.

Thus, the shredding of cotton stalks suitable for the production of composite wood-plastic board materials [63; p. 112-114] was tested on known equipment. At the same time, the modes and conditions of shredding of aggregates and machines were mainly changed without making changes in the design of these machines. All the above-mentioned shredding machines do not provide obtaining of conditioned wood-fibre pulp from cotton stalks that meet the requirements of wood-plastic board materials production [64; 27].

Scientists and employees of "KOMPOZIT NANOTEXNOLOGIYASI" Ltd. and scientists of the State Unitary Enterprise "Fan va tarakkiyot" of Tashkent State Technical University developed a device for shredding plant stems in bales [65; pp. 12-14, 66; pp. 22-24,]. However, this device also did not sufficiently ensure the production of conditioned chip masses from cotton stalks.

Studies conducted over a number of years on the possibility of shredding cotton stalks on existing serial equipment for shredding wood and various types of other plant raw materials have shown that none of the existing types of shredders is impossible to obtain a conditioned wood-fibre mass from cotton stalks, as they differ from wood and other wood waste by the presence of strong and elastic stem bark. The fibres of the bark envelop the cutting elements of shredding machines, which leads to frequent replacement and breakage of blades. At the same time, the size and shape of the resulting particles cannot be adjusted, and the shredded mass contains long bark fibres that have been soaked. Such an inhomogeneous mass is unsuitable for use, especially in the production of chipboards, especially wood-plastic board materials. As the quality of the pulverised mass in terms of particle size and shape is subject to high requirements.

On the basis of theoretical and experimental analysis of the known data it was concluded that it is necessary to develop a fundamentally new method and shredder of cotton stalks, providing cutting of the fibrous part (bark) of the stalk together with the woody part of a certain length with an even cut and obtaining woody-fibrous mass from cotton stalks.

§ 1.3 Study and analysis of the state of technological processes of manufacturing composite wood-plastic board materials from cotton stalks and polymer binders

Back in 1932, it was proposed to use the stems of annual plants to produce piezothermoplastic composite wood-plastic materials by the Barkley method [67; p. 4]. However, due to the complex technology, this method did not find practical application in industry. Later, a number of authors [68; p. 32-54] suggested using cotton stalks (guza-pai) as a wood filler for composite wood-plastic boards. Their recommendations were based on the experience in the production of composite wood-plastic boards from wood waste [69; p. 126], as well as annual plants such as sugarcane, flax, etc. The mentioned authors did not conduct experimental studies directly with guza-paya. It is necessary to note a very significant difference between guza-paya and such bast crops as flax, hemp, kenaf from wood - an additional operation of particle crushing is required, the content of bast particles in guza-paya is two times higher than their content in flax. These differences, as the attempts to obtain prototypes of boards from guza-paya in VNIINSM [70; p. 3] showed, did not allow to obtain either chips of optimal sizes or a homogeneous glued mixture. The strength values in one board fluctuated within large limits. High swelling of boards was also noted.

The works [71; p. 124-125, 72; p. 60] propose to produce boards from deregenerating wastes of annual plants, including guza-paya, using the reactivity of guza-paya components, first of all, lignin and carbohydrate part. Experimental studies in laboratory conditions have confirmed this possibility. However, it should be noted that due to the complexity of the technological process, manifested in the part of accuracy of maintaining the initial moisture content of the component, as well as to the size and shape of particles (all particles must pass through a sieve with a hole diameter of 3 mm, of which 60% must pass through a sieve with a hole diameter of 2 mm) [73; pp. 12-14, 74; p. 910], attempts to produce guza-paya board under production conditions at the Kherson pulp and paper mill (Morupinsk) of the Ukrainian Republic in 1975 were unsuccessful [75; p. 10-11].

The paper [76; p. 14-16] presents the results of research into the technology of production of guza-paya boards with the use of urea resins as a binder. The boards obtained in laboratory conditions were subjected to research on the fractional composition of raw materials, resin content, the influence of pressure and temperature of pressing and other technological factors.

The physical and mechanical characteristics of the boards have also been studied.

However, studies have been carried out in a limited range of filler particle sizes. Although in his time P.M. Dyskin's studies established the influence of particle shape and size on the quality of chipboards, the author of the above work limited himself to the fraction 10/5. P.M. Dyskin pointed out the growth of chipboard strength with the growth of chip length up to 40-50 mm [77; p. 18-23].

It should also be noted that coarse and fine fractions were not used in the manufacture of boards, while according to the author's data, when guza-paya was ground at moisture content of less than 15-20 %, the fine fraction was 32 %. Thus, it can be concluded that the proposed technology allowed only partial utilisation of cotton stalks, which is considered unacceptable. In addition, it should be noted that the results of laboratory studies were not brought to their production implementation.

The paper [78; p. 13-14] analysed the process of formation of wood-plastic boards from cotton stalks and presented the results of studies of the process of obtaining wood filler - guza-pai shavings. The influence of the fractional composition of chips and technological factors on the physical and mechanical characteristics of the obtained samples was studied. However, the author limited himself to a narrow range of chip length (up to the fraction 10/5) despite the fact that the strength of boards according to his own data increases with the growth of chip length [79; p. 32-35].

Both [80; p. 2] and [81; p. 110-142] did not investigate the degree of influence on the physical and mechanical characteristics of wood-plastic boards of the fibrous fraction, the separation of which, as will be shown in the following, is inexpedient and represents a difficult technological problem.

The results of the studies mentioned above have not found practical application, apparently, due to the complexity and energy consumption of the proposed technologies. Thus, in [82; pp. 90-128] the proposed technology includes washing of guza-paya under a ring shower at 25-30 °C with further soaking for 30 min in water at 50 °C, using also primary and secondary shredding, separation of chips into fractions and subsequent drying. Despite this, the cotton stalk is only partially utilised.

In [83; p. 202-262] it is proposed to introduce a new technological operation - freezing of cotton stalks with moisture content more than 30 % at temperature (-10,-15 °C), which also leads to additional energy consumption, lengthening of the technological cycle of stalks processing and its complication.

On the basis of specially conducted research in 1960-1967 in Uzbekistan [84; p. 82-88] a pilot batch of guza-paya slabs meeting the requirements of GOST 10632-63 grade PS-1 group B was obtained in production conditions.

These results were the basis for the production of chipboard wood-plastic boards from cotton stalks in the country [85; p. 101-142]. On the basis of the Decree of the Government of the Republic of Uzbekistan №-552 of 17 December 1969. [86; p. 4-15], a line for continuous pressing of multilayer chipboard K-175 of the company "Weite Metallwerk" was purchased from the FRG [87; p. 92-103] and installed in Sergeli town of Tashkent region. Preliminary laboratory studies have shown that with a slight change in the design of the units of the technological line it is possible to obtain crushed mass from cotton stalks, suitable for the production of composite wood-plastic boards [88; p. 10].

However, it should be noted that further production tests of the installed line K-175 showed that during the primary shredding of cotton stalks on the knife crusher of the company "Bison Pallmann" the blades of knives quickly become dull. As a result, the fibrous part of the stalk is not cut. The resulting long fibres wrap around the working parts of the machine, which eventually leads to clogging and breakage of machine parts. The presence of fibres in the stalk also disrupted the secondary chopping process. In addition, all subsequent units of the technological line are clogged with fibres [89; p. 28-30].

The specialists of the company concluded that on the line K-175 the production of boards is possible only from the wood part of the stem and proposed to separate bast fibres with the help of a separating plant of the company "Kastpun" The waste from separation, mainly bast fibres, making up to 40 % of the total mass, was proposed to be used for the production of cement-based boards [90; p. 8-21].

The technological process of cotton stalks processing proposed by the company "Bayte Metalwerk" is given in [91; p. 15-16]. The company's attempt to implement the proposed scheme was unsuccessful.

Tests of the cyclone and hopper complex built in Sergeli showed that the amount of wood particles in bast fibres was 22%, and the amount of bast inclusions, after the chips left the separation unit, was 24-25%   (Protocol of the results of the results of the

industrial tests of the separating plant for separation of bast inclusions from wood particles of cotton stalks from 27.10.79 in Sergeli). The firm also failed to solve the problem of further use of fibre [92; p. 102-189].

It should be noted that the separation plant created a strong dustiness in the surrounding area, as a result of which the operating personnel had to work in respirators. It should be added that even for this method, the cotton stalks have to undergo a primary crushing stage, as it is practically impossible to separate the stalks' crust by mechanical means.

The issues of grinding and development of necessary parameters of technological process for obtaining boards from cotton stalks with a

taking into account their specificity. Therefore, the production of particleboards from guza-pai has not been established so far. Line K-175 is used for production of particleboards from wood waste.

The paper [93; pp. 61-65] presents a scheme for the production of boards from cotton stalks proposed by Cofram Siempelkanp. This company, like Bahre Metalwerk, proposes to use only the wood part of cotton stalks, and to exclude bast fibres and dust-like fraction from the process. It must be assumed that the company is trying to use the existing standard equipment for the production of boards from wood chips. It should be noted that separation of bast fibres from stems is a complicated, expensive and dusty process, and 40-50% of the guza-paya mass is not used in production. So, we can conclude that the existence of numerous studies and experience in the production of wood chipboard is still insufficient. For the successful production of board materials made of guza-paja, further special research was necessary, taking into account the specificity of the material [94; p. 201-238].

Thus, the conducted review of researches on obtaining composite wood-plastic materials and boards from guza'pai and polymer'r binders showed the following:

In pre'.dchozhe'nye earlier technologies of obtaining wood-plastic boards from guza-paya (except for the work [95; p. 3-8]) only the wood part is used, i.e. 50-60 % of the total mass of cotton:

To date, none of the proposed technologies has found industrial implementation due to the complexity and incompleteness of the technology of grinding cotton stalks and obtaining conditioned wood fibre pulp.

There are no sufficiently clear theoretically substantiated and experimentally confirmed data on the rational size of wood particles, depending on the requirements to their strength characteristics. The results of the researches carried out in this field are contradictory, moreover, in [96; p. 8-18] the optimum sizes of wood shavings are determined for boards made not only of guza-paya, but with the addition of wood shavings. In [97; p. 11] and [98; p. 24] the research was carried out using guza-paya wood chips without its fibrous part and fine fraction, which also limits the scope of application of the obtained results in the production of wood-plastic boards.

It should be noted that the above-mentioned authors have not carried out studies on the influence on the physical and mechanical properties of wood-plastic boards from [99; p. 30-32] guza-pai with the content of bast fibre, and studies on the influence of fine fraction have been carried out in a limited volume, which did not allow the authors to give sound recommendations, and the conclusions of the conducted studies to determine the optimal [100; p. 52-56] percentage of binder content are true only for the investigated filler fractional composition, i.e. without the use of fine fraction and in the absence of free bast fibre. Therefore, it is necessary to carry out in-depth scientific research in the field of development of a new method and creation of a device for grinding and production of wood fibre filler from cotton stalks [101; p. 30-32].

Thus, in order to obtain composite board materials with high physical and mechanical properties from cotton stalks, it is necessary to develop a new method and device for grinding cotton stalks, allowing to obtain a conditioned wood-fibre mass from them and, accordingly, composite boards of high quality. To fulfil the set goal it is necessary to solve the following tasks:

development of technology of cotton stalks grinding in order to obtain wood-fibre filler for composite wood-plastic materials;

development of a method and improvement of machines and installations for grinding, reconstruction of known and creation of new products and accordingly technological line of grinding and production of wood-fibre filler from cotton stalks;

development of technology for obtaining composite wood-plastic board materials using the created wood-fibre masses from cotton stalks and polymer binders [102; p. 48-52];

research and development of a scientifically substantiated approach to the possibility of obtaining composite board materials for special purposes based on wood-fibre fillers from cotton stalks and polymer binders,   in particular with decorative coatings in the

furniture industry, construction industry and mechanical engineering [103; p. 3233].

§1.4 Conclusions to the first chapter

1.   The current state of the art of annual plant stems has been considered in detail and the possibility of using them in the development of composite wood-plastic materials based on them and polymer binders has been analysed.

2.   Studies and comprehensive analysis of literature sources of existing methods and devices for grinding woody and wood-like stems of annual plants, including cotton stems in order to obtain wood-fibre pulp from them and wood-plastic board materials accordingly are given.

3. On the basis of theoretical and experimental works it can be noted that for obtaining composite board materials from cotton stalks it is necessary to develop a fundamentally new method of grinding, which provides obtaining conditioned mass from cotton stalks.

4. The study and analysis of the state of technological processes of manufacturing composite wood-plastic board materials from cotton stalks and polymer binders is considered .

5. The goals and objectives of further research are formulated, including in the field of the method and installation for grinding cotton stalks and obtaining from them a conditioned wood-fibre pulp for the production of wood-plastic board materials.

CHAPTER II

SELECTION OF OBJECTS, METHODOLOGY OF OBTAINING AND RESEARCH OF PROPERTIES OF WOOD-PLASTIC BOARDS MATERIALS

§ 2.1 Selection and justification of research objects and their characteristics

We have chosen cotton stalks for obtaining wood fillers. Cotton stalks among annually renewable annual plants- kenaf, rice, sunflower husk, ground nuts-have more woody part and it is comparatively convenient to grind them.

Let us consider cotton stalks as a material for chipboard production. As already mentioned, they are closer to wood in chemical composition, structure, and properties compared to other annual plants [104; p. 42-44]. Table 2.1 summarises the main physical and mechanical characteristics of woody species and cotton stems [105; p. 32-34].

Table 2.1 **Main physical-mechanical properties of wood species used in the production of wood-plastic boards, and cotton stalks**

Wood species	Volumetric weight in completely dry state, g/cm^3	Static bending strength, MPa
Pine	0,39-0,46	64,9-87,7
Spruce	0,39-0,46	60,3-77,4
Aspen	0,39-0,47	58,0-76,6
Cotton stalks	0,38-0,40	60,0-88,0

As can be seen from Table 2.1, the volumetric weight and static bending strength of cotton stalk are almost similar to those of wood species used in the production of wood-plastic boards [106; p. 8-9].

It is known that the acidity (pH) of wood influences the curing process of the polymer resin binder and, consequently, the properties of wood-plastic boards. The pH values of wood of different species vary between 3.3-5.15. The woody part of cottonwood stem has a pH of 3.2-5.10. This also indicates the proximity of cottonwood properties to those of wood [107; p. 18-19].

It is also known that the best raw material for composite wood-plastic boards is light wood with a volume weight of up to 500 kg/m^3 [108; p. 9]. The volumetric weight of cotton stalks is 380-400 kg/m^3 , which also indicates the possibility of replacing wood raw materials with cotton stalks [109; p. 28-30].

Unlike wood, which is used after pre-barking, cotton stalks are used together with the bark, as the process of separating the bark before chopping is not practically feasible. In addition, the bark makes up, as already mentioned, one third of the total volume of the stalks and its separation is not economically feasible.

Let us consider the constituent parts of the cotton stalk that form the chip mass when crushed.

The bark is 0.3-0.4 mm thick, tightly adhering to the woody part of the stem, and consists of tough bast fibres. The fibres are cross-linked and separate from the stem in the form of strips. When the bark is rubbed, the upper peel and the transverse bonds are broken and the bast fibres can be clearly distinguished. They are very elastic, flexible and have a high tensile strength. They are the main reason for the high strength properties of the stem.

The woody part of the stem makes up about 60 per cent of the wood. It is characterised by high mechanical properties and produces high quality chips when milled. Its structure is close to that of light wood. A distinctive feature of cotton stalks from wood is the presence of up to 5 per cent of the core, which consists of cell parenchyma. The degree of moisture also has a significant influence on the process of stem crushing [110; p. 6]. It affects the size and shape of particles, energy consumption, cutting tool life, etc. Based on the study of these factors, it was found that the optimal moisture content of cotton stalks is 30-35 %. However, it should be taken into account that each method of shredding is conditioned by the choice of optimal moisture content and it should be determined depending on the applied shredding technology [111; p. 2829].

The properties of cotton stalks change significantly depending on the period and method of storage. Under the influence of light, oxygen, air temperature, precipitation (in open storage) and living microorganisms, cotton stalks undergo component changes.

Thus, cotton stalks of Tashkent-1 variety, storage period up to 1 year, were selected as an object of research for obtaining wood filler.

Urea-formaldehyde resin of KFMT grade (GOST 14231-90), viscosity 4060 s, was used as a polymer binder. When conducting a special experiment to determine the quality of composite wood-plastic material, the concentration of urea resin varied, both downward and upward, from this indicator.

§2.2 Methodology
for obtaining samples of wood-plastic
composite board materials for determining their physical and
mechanical properties and factors affecting them

Taking into account that in our research the principle used in the production technology of guza-paya boards is based on the principle used in the production of particleboards, when analysing the technological factors we used the information on particleboards (particleboard) in an improved form, as applied to wood-plastic composite board materials [112; p. 84].

The factors influencing the formation of physical and mechanical properties were divided into the following groups:

1. Factors characterising the filler:

structure and properties of cotton stalks;

Structure of cotton stalk chip mass;

particle size and fractional composition of the chip mass.

2. Factors characterising the technological process:

moisture content of the filler before and after sintering;

amount of binder;

plate design;

slab density;

pressing mode.

Since unlike the known and well-studied particleboard technologies, this paper uses a non-traditional material - cotton stalks, we proceeded from this premise when determining the factors to be studied. The distinctive feature of guza-paya from wood is mainly in the physical structure of the stalk. The shredded mass of guza-paya contains about 21-24%   fibre content

inclusions formed as a result of bark crushing, which is not the case with

particleboard. This changes the properties of the filler to a certain extent.

In addition, from the analysis of the results of the studies carried out so far, it was found that it is the individual features of the filler from cotton stalks were not paid attention to in the development of the technology of chipboard from cotton stalks [113; p. 30-32]. In this connection, we have studied in detail the structure and properties of cotton stalks, fractional composition of the crushed mass: particle geometry, friability, volatility, bulk density of guza-pai chips, etc. [114; c. 4-5].

Constant factors. Some factors were assumed to be conditionally constant, these are mainly factors not related to filler properties such as, board design, type of binder, pressing mode.

Slab design. It is of great importance, as it mainly determines the conditions of heat and mass transfer, physical-mechanical and operational properties of the boards. In a multilayer board, larger particles are placed in the middle of the package, while smaller particles make up the outer layers. This allows them to be compacted more tightly, which in turn increases the static bending strength of the board.

Increasing the binder content in the outer layers intensifies the heat and energy transfer to the middle of the board due to moisture and pressure gradients generated in the board during pressing.

A single-layer board has more uniform properties throughout the volume, as the particle size and binder content are the same in all directions. This results in a more uniform density over the thickness and therefore a higher tensile strength perpendicular to the plate.

Considering that the difference in particle size is large in the ground guza-pay, the present study also decided to develop a technology for obtaining multilayer composite wood-plastic board materials [115; p. 20].

Type of binder. Urea-formaldehyde resins are mainly used as a polymer binder in the production of chipboard. They are colourless and odourless. The Central Scientific Research Institute of Woodworking Industry has developed a formulation of non-toxic urea-formaldehyde resins containing 0.2-0.3 % of free formaldehyde, which meet the requirements of sanitation and hygiene. In this work urea-formaldehyde resin of KF-MT grade of Navoi nitrogen JSC according to TSh 6.1-00203849-104:2007 was used.

It is known that the temperature, time and pressure of pressing have a significant influence on the properties of the boards. However, in this work it was not aimed to study the pressing process in the aspect of broad chemical and thermophysical phenomena accompanying it. The pressing regime corresponded to the generally accepted parameters of the technology for obtaining particleboard based on urea resins and similar technologies for other materials, in particular, cotton stalks.

Variable factors. The paper analyses the influence of a number of factors and their arrangement by degree of influence on the physical and mechanical properties of particleboard. As the most significant factors were established: the amount of binder, density of the board, fractional composition of the filler. This series also includes the moisture content of chips before and after grinding [116; p. 21].

Amount of binder. The bonding strength of the filler particles depends on the formation of a glue joint between them, which should not have significant internal stresses due to glue shrinkage. As the adhesive interlayer thickens, the internal stress increases, resulting in a weakening of the adhesive bond between the adhesive and the filler particles. 34 Therefore, theoretically, a single molecule thick adhesive layer, i.e. a monomolecular layer, should have the maximum strength. On the other hand, insufficient osmolisation of the particles leads to a decrease in the physical and mechanical properties of the board material.

In particleboard production, when using urea-formaldehyde adhesives, it is recommended that at least 4-3% of the binder be added to the chip mass. The maximum amount of binder is 14-20%.

It is known that the amount of binder depends largely on the raw material used, i.e. wood species, chip quality, particle shape and size.

The structure and composition of the pulverised mass of cotton stalks require the establishment of a binder rate for boards. For example, cotton stalks contain about 33% of   bark, which has more bark after shredding.

adsorption capacity compared to wood. The size and geometry of cotton stalk particles have a wide variation of parameters. Particles with irregular rough surface prevail in the cossettes, which predetermines an increase in the amount of binder.

Density of the board. Analysis of literature sources and results of experimental studies show that density has a decisive influence on the physical and mechanical properties of composite boards.

Due to the fact that it is impossible to obtain slabs with a constant density during the experiment, it was necessary to take into account the density of each sample when assessing its quality. It is known that an increase in density leads to a reduction of voids in the slab, which makes it acquire high strength properties and water resistance. However, an excessive increase in density leads to destructive phenomena in the volume of the slab and an increase in its volumetric weight.

The density of the board varies depending on the density of the filler material. The most dense boards are made of coniferous wood, which has a high specific gravity. Light woods, as well as cotton stalks close to it in density, can have a medium density. In all cases where a previously unknown composition is used, it is necessary to determine the effect of density on material properties by experimentation.

Fractional composition of the filler. This indicator is characterised by the content of fibrous inclusions in the ground mass, in the form of bast bark, formed during the grinding of wood parts, as well as dust-like particles formed from the core of the stem. We have taken into account that the fractional composition is based on the results of sieve analysis of the pulverised mass, dividing it into several parts, especially in terms of particle size.

The influence of fibre inclusions is obvious, as fibres act as a reinforcing and bonding element. At the same time, they require a change in the sintering and pressing regimes of the material, which has a great influence on its properties. The same is true for the wood and dusty part of the chips, hence the determination of the optimum ratio of fractions should be an integral part of the study.

The particle size of the filler is of great importance in the creation of composite material, as the quality of the chips is determined by the content of particles of the right size in the chips. Determination of the optimum filler size is also important for establishing the cotton stalk shredding regime.

Moisture content of the cossettes before and after bleaching. The moisture content of the cossettes before tarring is mainly formed by the natural moisture content of the cotton stalks and has an influence on the tarring process. It is known that the more the wood cells are saturated with moisture, the less resin they absorb. It remains on the surface of the particle and a continuous layer of glue is created there. At the same time, excessive moisture leads to delamination of the board after the pressing pressure is removed.

The moisture content of the chip mat after flaking includes the natural moisture of the chips and the water contained in the binder. The amount of moisture and its distribution over the thickness of the formed mat has a strong influence on the properties of the finished board. Lack of moisture leads to reduced heat transfer and poor internal particle bonding, while excess leads to slab relaxation due to high vapour-gas pressure when the slab is removed from the press.

§ 2.3 Methodology for obtaining composite wood-plastic
board materials based on wood fillers from
cotton
stalks
and polymer binders

The stages of the technological process of wood particleboard production are described in detail in [117; p. 20-22]. Based on the technological processes of the mentioned works, as well as the factors developed by us [118; p. 65-71], influencing the formation and magnitude of physical and mechanical properties of wood-plastic boards, we determined the methods of their production [119;p. 108-162].

Stems of medium-fibre cotton (cotton) of one-year storage were used for research. Cotton stalks were fed for shredding into the primary guza-pai shredder according to the method developed in the State Unitary Enterprise "Fan va tarakkiyot" [120; p. 142-

165] [120; c. 142-165]. The scheme of modernised plants for obtaining chips and shavings from cotton stalks is given in the appendix of the thesis [121; p. 82-97].

The particle sizes were 5-100 mm, with a weighted average size of 17.3 mm and a free fibre content of 12-10 %. The particles obtained after primary shredding with a moisture content of 10-14 % were fed for secondary shredding into a laboratory shredder with a knife shaft. The resulting shavings were needle-shaped with a flexibility of 1/5   +1/25.

Fractional composition of chips is given in Table 2.2.

Table 2.2.

Fractional composition of shavings obtained in a laboratory shredder based on a modernised VT-31 machine tool

Fractions	$\frac{10}{5}$	$\frac{5}{2,5}$	$\frac{2,5}{1,5}$	15 1,0	10 0,63	0,63_0
Fractional content to total mass,/ %	26,8	17,9	20	18	10,3	7

Chips consist of four fractions: coarse fraction 30/10, medium fraction 10/1.5, fine fraction 1.5/0.63, dust fraction 0.63/0. Each fraction was stored separately and used in obtaining the package in the calculated amount depending on the belonging to the layers according to Table 2.3.

The chip pack was moulded in three layers. The storage, threshing and moulding of the outer and inner layers were performed separately. In order to obtain uniform moisture content throughout the whole mass, the chips were stored for 3 days before drying in chambers with air humidity of 75 ±3%.

Table 2.3.

<u>**Distribution of fractions on the outer and inner layers of the package**</u>

Indicator	Inner layer				Outer layer			
Fraction	$\frac{30}{10}$	$\frac{10}{1,5}$	$\frac{15}{0,63}$	0,63_0	$\frac{30}{10}$	$\frac{10}{1,5}$	15 0,63	0,63_0
Content to total weight	30	30	5	5	0	0	20	10

^Note: numerator is chip length, denominator is chip thickness.

The air humidity was monitored with a psychrometer at room temperature. After 3 days of storage the humidity of the chips was 14 %. The chips with the calculated amount of the corresponding fractions were placed in a cloth tray for drying.

Drying of chips was carried out in an electric dryer, where heating of chip material was carried out by forced supply of hot air through the mass of chips coming from an electric heater. The drying temperature reached 140°C. The humidity of chip material after drying was measured by weight method on the moisture meter VLB-100.

All components were calculated on the basis of the required number of absolutely dry chips, which is determined by formula (2.3.1) [122; p. 8297].

$$G_0 \quad \frac{G_n}{(100+W_n)*(100+P)} \cdot \frac{}{10^{4}*y*_H} \qquad (2.1)$$

where y is the given density of the slab, y = 0.7 g/cm^3 ;

and - volume of uncut slab (after pressing), cm^3 ;

W_n - moisture content of finished boards (it is taken as 8 %).

Characteristics of KFMT resin and its physical and mechanical properties are given in Table 2.4.

For our experiment, the volume of the slab was:

$$V = 28.4 \times 47.0 \times 1.6 = 2135.7 \text{ cm}^3 ,$$

where W_n - moisture content of finished boards (it is taken as 8 %)

P - binder consumption rate (to the mass of absolutely dry wood) - 10 %. Substituting the values into the formula (2.3.1), the absolutely dry weight of the chip mass was determined. $Go = 1282.8$ g.

Table 2.4.

Indicators of physical and mechanical properties of KFMT resin

Mass fraction of dry residue, %	66
Viscosity by viscometer VZ-4 at 20±0,5 °C,	45
Hydrogen ion concentration, pH	7,5
Curing time at addition of 1% - at 100 °C	40

For the experiment, the mass ratio of the inner and outer layers was assumed to be 70 and 30%. Based on this, the masses are respectively: $G_o^H = 880.8$ g, $G_o^H = 377.3$ g. The total weight of chips, (Gw) consumed per board at given values of chip moisture content can be determined by the formula:

$$\frac{10^2 \cdot y \cdot u \cdot (100+W_{CTp})}{(100+W_n) \cdot (100+P)} \qquad (2.2)$$

where *Wcmp is the* specified chip moisture content before osmolisation. Total weight according to the adopted mass ratio is distributed over inner and outer layers, the results of calculations are given in Table 2.4

The amount of binder is calculated in relation to the mass of absolutely dry chips. The binder consumption (C)$_o$) (by dry residue) per board is determined by the formula:

$$\text{P}^H = 12\%; \text{P}^{BH} = 9\% \text{ тогда}$$
$$Qo = G_o \cdot P/\text{ЮО} \qquad Q_o = 45.3 \text{ г}; Q^{BH} = 79.2 \text{ г} \qquad (2.3)$$

Due to the fact that a water-colloidal solution is used as a binder, the consumption of liquid binder (C)$_ж$), per one board, was determined by the formula:

$$Q. = Q.0 \cdot \wedge, \qquad (2.4)$$
$$1\backslash$$

where K is the dry matter content of the binder. For our experiment K = 54 %.

The amount of hardener for the outer and inner layers was assumed to be 5 % of the liquid resin weight. The hardener for the outer layer consists of 20 % ammonium chloride and 30-25 % ammonia solution in water. Ammonia for the outer layer is added to prevent premature curing. The hardener mass for the inner layer consists of 20 % ammonium chloride and 80 % water.

Mixing of wood filler (chip material) with binder was carried out on a laboratory cyclic mixer. The mixing time was 3 min. The rotation frequency of the mixer shaft was 240 rpm. Osmolised particles of wood filler were aged for 20 min. Osmolised material was placed on a special frame located on a metal pallet made of stainless steel. The frame has internal dimensions of 284x470x170 mm and contributes to the formation of a three-layer composite wood-plastic board (Table 2.5).

Table 2.5.

Estimated number of chips per plate

Package Layers	Fractional composition	Fractional content to total package weight	Weight of chip filler, g.
Internal	30/10	30	334,8
	10/1,5	30	384,8
	1,5/0,63	5	64,1
	0,63/0	5	64,1
Outside	1,5/0,63	20	256,6
	0,63/0	5	123,3
Total weight of the package, g		100	1282,8

Hot pressing was carried out on a hydraulic press with a
with built-in thermostatically controlled heaters.

Depressurisation was carried out within 1 minute, stepwise. At that, the pressure reduction of the first stage was % of the maximum pressure. The duration of pressure reduction of the last stage was equal to 10 s. The finished plates were kept at room temperature (20-25^0 C), and air humidity 65±5 % for 5 days, only after that the specimens were cut out for testing.

We will prepare samples from wood-plastic board materials for the study of their physical and mechanical properties in accordance with the methodological grid [122; p. 3-15, 82-104].

§ 2.4 Methodology for determining physical and mechanical properties of wood-plastic board samples

In accordance with GOSTs adopted and valid for the CIS states [123; p. 12], the percentage content and physicomechanical properties of composite wood-plastic board materials [124; p. 30-37] and their pressing modes were selected for chip evaluation. These include the following indices [125; pp. 5-16,   200-209]: tensile strength at static bending [126; p. 3-10, 91-102], tensile strength perpendicular to the plate, specific resistance to pulling nails out of the plate, specific resistance to pulling screws out of the plate, water absorption and swelling in thickness for 24 hours. The density of the board was measured using known methodology. The manufactured composite wood-plastic boards had the following dimensions: 450x450x16 mm [127; p. 366].

In order to control the quality of the slabs, they were divided into samples according to the scheme shown in Fig. 2.1, which sufficiently ensures that samples are taken from different areas of the slab.

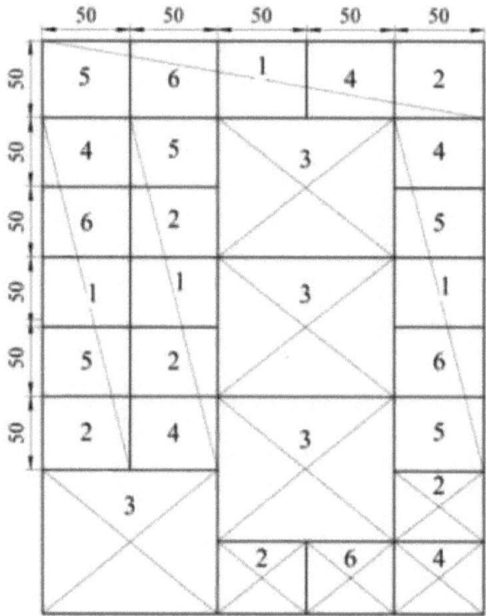

1 - for bending and modulus of elasticity; 2 - for stretching perpendicular to the plate;
3 - water absorption and swelling; 4 - screw pulling;
5 -
nail pulling; 6 - hardness.

Fig. 2.1. Schematic of cutting out prototypes for physical-mechanical tests

It should also be noted that from the finished samples of composite wood-plastic boards it is envisaged to cut out edge zones with a width of at least 15 mm, which have a less dense and unstable structure. At the same time for each type of tests one sample was taken close to the centre and the other from the edges of the board, which allows to reduce the error of the experiment associated with the location of the test sample on the plane of the board, i.e. their standard dimensions corresponded to GOST 10633 "Particleboard".

The general rules of preparation and carrying out of physical and mechanical tests provide for sampling at a distance of 150 mm from the transverse edge of the board. In this research work, due to the limited capabilities of the technological equipment available in the State Unitary Enterprise "Fan va Tarakkiyot", as well as in order to save materials and labour costs, composite wood-plastic board samples were taken with reduced dimensions, which allowed to use the edge zones for testing.

Since the density of the latter differed from that of the central zones, the results of measurements of the above-mentioned slab quality indicators for the edge and central zones were processed separately.

The physical and mechanical properties of composite wood-plastic [128; pp. 3-10, 102-154] boards were determined according to the following

listed: GOST 10633 "Particleboards", "General rules for preparing and conducting physical and mechanical tests"; GOST 10635 "Particleboards. Methods of determination of strength and modulus of elasticity in bending"; GOST 10636 "Particleboards. Methods of determination of tensile strength perpendicular to the plate plate"; GOST 10637 "Particleboards. Method of determination of specific resistance to pulling out of nails and screws"; GOST 11843 "Particleboards. Method of hardness determination". Detailed descriptions are given in [129; p. 156-163].

Density determination. The density of the board was determined according to GOST 1063490. The thickness of the sample was measured in four points with an accuracy of 0.01 mm. The arithmetic mean of four measurements was taken as the thickness of the sample [130; p. 67-73]. Then the samples were weighed to the nearest 0.01 g. Density was determined to the nearest 0.01 g/cm^3 using the formula:

$$p = m/b \cdot I \cdot h \tag{2.5}$$

where: m - sample mass, kg (g);

1 - sample length, m (cm);

b - width of the sample, m (cm);

h - thickness of the sample, m (cm).

Determination of water absorption and swelling. Water absorption for 24 hours was determined according to GOST 10634-90. The samples were immersed in water to a depth of 20 mm. The water temperature was 290+2 K. Before measuring and weighing the samples were dried with filter paper.

Water absorption was determined to the nearest 1% using the formula:

$$Д\ W_{Bg} = 1\,0\,0 \cdot (m_1 - m)/m, \tag{2.6}$$

where m - mass of the sample before moistening, kg (g); mi - mass of the sample after moistening, kg (g).

Linear swelling was determined with an accuracy of 1% according to the formula:

$$Д\,h = 100\,(h1\text{-}h)\,/\,h, \tag{2.7}$$

Where h- thickness of the sample before moistening, cm (mm),

hi - thickness of the sample, after moistening, cm (mm)

Determination of tensile strength and modulus of elasticity in static bending. The bending strength was determined according to GOST 10635-90. The distance between the supports was 200 mm. The test was carried out on a universal testing machine UM-5.

The bending strength was calculated in MPa by the formula:

$$C_{изг} = 3\,PI/2\,\text{b}\,\text{h}^2 \tag{2.8}$$

where P is the load acting on the specimen at the moment of fracture, N;

l - distance between the supports of the testing machine, m (cm) b - width of the specimen, m (cm);

h - height of the sample, m (cm);

The bending modulus of elasticity Ei was calculated by the formula:

$$^\wedge(F._-FJ$$
$$^u\ 4bh^3(S_2-Si)$$, (2.9)

where - distance between the supports of the testing machine, m (cm);

b - specimen width, m (cm); h - specimen thickness, m (cm);

F2-F1 - load increments on the rectilinear section of the graph of deflection dependence on load, determined with an error of not more than 1 %, N.

S2-S1 - increase in deflection, determined with an error not exceeding 0.01 mm (cm).

Hardness determination. Hardness of the boards was determined according to GOST 11843, for which samples were taken according to GOST 10633-90 in the size of 50x 50x 16 mm.

The hardness of the plates was determined at the point of intersection of the sample diagonals. The specimen was placed in a special fixture, the ball was placed on the specimen and the rod with the plate was lowered, the indicator arrow was set to the zero position.

The fixture with the specimen was placed in the UM-5 testing machine and immersed at a rate of 10 mm/min until the ball reached an indentation depth of 2.0+0.05 mm. At this point, the load was measured with an error of not more than 10N.

Hardness was calculated using the formula:

(2-10) where P is the load when the ball is pushed into the specimen to a depth of 2.0 mm, N;

F = footprint projection area, m^2 (see).^{2}

4

Determination of tensile strength perpendicular to the plate plate. This indicator was determined according to GOST 10633-90.

Samples were taken with the size of 50x50x16 mm. The specimens were glued on both sides with the adhesive onlays prepared from KFMT resin. Then the specimens were etched into a special fixture on the UM-5 testing machine. The tensile strength perpendicular to the plate plate was calculated according to the formula;

$$tfp=£$$. (2.11)

where 1 - length of the specimen, m (cm);, b - width of the specimen, m (cm) P - *the* greatest load acting on the specimen at the moment of fracture, N.

Determination of specific resistance to pulling out nails and screws. Samples are taken according to GOST 10633. The sample size is 50x50x6 mm. The nail was driven into the layer of the slab to the thickness of the slab. Screws were screwed into pre-drilled holes of 2 mm in the sample to the thickness of the board.

The nails and screws were pulled out in the direction of their axis at a speed of 10 mm/min of the travelling gripper of the testing machine.

The specific resistance to nail pulling was calculated in MPa using the formula:

$$p \qquad (2.12)$$

where P_{max} is the highest load, N;

d - diameter of nails, m (cm); P - *length of the* driven part of the nail, m (cm);

The specific resistance to screw pulling was calculated in N/m using the formula:

$$\text{ч} \qquad p \qquad (2.13)$$
$$\qquad \text{ь}$$

where P_{max} *is the* highest load, $N;$

/ - depth to which the screw is screwed into the sample, (mm).

§ 2.5 Mathematical and statistical processing of experimental scientific research results

The methods of processing the results of measurements with multiple observations are described in the standards of the State System of Uniformity of Measurements and in the works of metrological institutes of the country as well as in [131; p. 3-8, 20-48].

To evaluate the quality of the tested composite wood-plastic boards, the following statistical values were calculated for each indicator.

Arithmetic meanDx) by formula

$$x = \_Sja^{x_J}x_J, \qquad (2.14)$$

standard deviation (S) according to the formula

$$\cdot = -T_{i^J}\left(x_i + x\right)^2 = \Gamma_{\pi x} 1^{x^2} \pi) - ;\circledS = i^{x}J^{*}j)^2\right] > \qquad (2.15)$$

Mean error of the arithmetic mean (t) according to the formula

$$m = \quad +\frac{s}{\sqrt{n}} \qquad (2.16)$$

accuracy index (P) in per cent according to the formula

$$\qquad (2-17)$$

where xj *is the* value of the characteristic in the middle of the interval;

n is the number of observations enclosed in) -th interval;

e is the number of intervals.

The confidence limits for the mean parameter at the significance level q were determined from the ratio

$$X^-u,^\wedge < Xo < X + ^\wedge \qquad (2.18)$$

where $t/\ k$ is the Student's coefficient for significance level a and number of degrees of freedom $k = n - 1$.

The research used a 5% significance level accepted in the industry.

The confidence interval for the mean square deviation was calculated by the formula

$$S\,Z_1 \sqrt{\frac{n-1}{n}} \overset{< Gf < 5\ Z_2^\wedge}{} \ \{\Pi \qquad (2.19)$$

where Z_i and Z_2 are coefficients depending on the confidence level and the number of degrees of freedom.

According to GOST 11.002, the abnormality of the observation results was assessed. Since this task is solved by the method of statistical evaluation, our decision about the abnormality of the observation result may be wrong with a certain small probability. Therefore, in such cases, first of all, the conditions of the experiment were analysed and the reasons for the sharp deviation of the observation results were identified, and if possible, the measurements were repeated.

In other cases, the abnormality of the observation result was assessed according to the criterion of N.V. Smirnov [132; pp. 201-228].

For a number of measurement results of one and the same sample $x_1 < x_2 \cdots < x_n$ arithmetic mean x *was* determined by the formula, mean square deviation by the formula

$$4^* = \qquad И \, Л_{X_n} = (^\wedge \qquad\qquad (2.20)$$

Using D $_{xi}$, D $_{xp}$, *a* given probability P_{zad}, *the* number of observations n and a table of limit values *in the* case of unknown general mean square deviation solved the question of abnormality of the results of observations and $x_1 \, u \, x_n.$

When $(x_n - x)/S > 6$ or $(x - x_1)/S > 6$ result respectively xp *and* x^1 was considered abnormal.

The Romanowsky criteria were used to decide whether the difference $2 \, ^2$ between the estimates < and was significant or random.

At the same time we calculated the values

$$_{=тт} F^{15} \rangle\!\rangle = J^Z \S H F'^R {}_{u=2}B \qquad\qquad (2.2^1)$$

where v_1 and v_2 *are the* numbers of degrees of freedom, respectively, for and $S'_1, S'_1; F = S_f S^\wedge$

If $R < 3,$ then with a probability greater than 0.889, we can say that the discrepancy between the S_1, S^1 estimates is random.

Calculation of the amount of fillers based on cotton stalks and added polymer resin. All components were calculated based on the mass of the slab.

The weight of the slab is:

where a - slab length, cm; b - slab width, cm; c - *slab* thickness, cm; y - *slab* density, g/cm^3 .

$$m_u = a \, b \, c \, j, \qquad\qquad (2.22)$$

To calculate the weight of the slab, a sample of 42 x 42 x 1.6 cm with a density of 0.70 g/cm^3 was taken. The weight of the slab sample is

$$m = 42x \, 42x \, 1.6 \, x0.7 = 1976.$$

Percentage ratio of inner and outer layers 50 %: 50%.

On this basis, the mass of the inner and outer layers is 988 g.

Knowing the mass of the binder we calculate the mass of absolutely dry chips.

Weight of absolutely dry chips

$$m^\circ = {}^I o\!\sim = {}_{\text{АТ}\text{М}} = 9\ 0\ 0\ \text{г.,} \tag{2.23}$$

where W - moisture content of ground guza-paya; W=4 %
Amount of finished resin solution:

$$Q_w = mo\ _w\ P/K = 950\text{-}12{:}55{=}207{,}3\ \textit{г,}$$
$$Q_{BR} = m_{о\ вн}\ P/K = 950\text{-}10{:}55{=}172{,}7\ \textit{г,} \tag{2.24}$$

where P - binder consumption rate in relation to the mass of absolutely dry chips, %;
K - resin concentration, %; $P_{\text{нар}} =12$ %;$P_{\text{внут}}$ p=10%;K = 55%
The amount of hardener for outer and inner layers is 55 % of the resin weight

$$Onar = Q_{mP} - 0.05 = 207.3 - 0.05 = 10.4\ g,$$
$$Ovn = Q_{BR} - 0.05 = 172.7 - 0.05 = 8.6\ g.$$

The hardener for the outer layer consists of 20 % ammonium chloride, 25+30 % of 25 % ammonia solution and 50 - 55 % of water $\text{tkn}_\text{ч}$ cg= 0,2 - 10,4= 2,1 g;

$$m_{\text{NH}_q\text{OH}} = {}^{0,3\text{-}10,4\underline{=}3,1\ \textit{г}}$$

$$T_{\text{H2O}} = 0{,}5 - 10{,}4 = 5{,}2\ \textit{г.}$$

Ammonia for the outer layer is added to prevent premature curing.
The hardener for the inner layer consists of 20% ammonium chloride and 80% water:

$$m_{\text{NH}_q\text{EL}} = 0{,}2 \cdot 8{,}6 = 1{,}7\ \textit{г;}\ \ ш_{\text{H2O}} = 6{,}9\ \textit{г.}$$

§ 2.6 Conclusions to the second chapter

1. Research objects for wood-plastic board materials were studied and analysed.
2. Modern methods and instruments were used in the study of physicochemical and mechanical properties of wood-plastic board composite materials.
3. Methods of obtaining composite wood-plastic board materials based on wood fillers from the stems of annual plants and polymer binders are considered.
4. A methodology for determining the physical and mechanical properties of wood-plastic board materials has been carried out.
5. The methods of mathematical and statistical processing of the obtained experimental scientific results are described.

CHAPTER III

STUDY OF COMPOSITION, PHYSICO-CHEMICAL AND MECHANICAL CHARACTERISTICS OF COTTON STALKS AND DEVELOPMENT OF A METHOD OF GRINDING AND PRODUCTION OF WOOD-FIBRE PULP
ON THEIR BASIS

§ 3.1 Study of the composition, physicochemical and mechanical characteristics of cotton stalks for the production of conditioned wood-fibre pulp for the production of wood-plastic board materials

Cotton is an annual field plant [133; p. 19]. In terms of natural properties, cultivated varieties differ significantly from wild varieties. Cotton grows in the form of bushes. The plant height varies widely (0.8: 1.4 m) depending on the variety and agrotechnical methods of cultivation. Side branches and fruit-bearing branches with bolls come from the base of the stem. The bark is 0.3-0.4 mm thick, tightly adhering to the woody part of the stem and consists of tough bast fibres. The fibres have transverse bonds and are separated from the stem in the form of strips. In cotton stalks with moisture or after their long storage, when the bark is rubbed, the upper skin and transverse bonds are destroyed and the bast fibres can be clearly distinguished with the naked eye. They are very elastic, flexible and have a high tensile strength. To a greater extent, they are responsible for the high strength properties of the stem [134;p. 112-148].

The bolls on the cotton stalks remain in very small numbers after harvesting by the cotton picker. Thus,

the ratio of individual parts of the stem weight is in (%) [135; p. 736741]:

stem wood - 50

boxes - 6

bark-30

root part - 14

All cotton organs are valuable raw materials for industry and can be fully utilised. Various products can be obtained from them [136; p. 737-743].

In particular, cotton stalks can be used to produce hydrolytic alcohol, fodder yeast, and roughage.

When developing composite board materials, the structure, chemical composition and physical and mechanical properties of the filler are of great importance. Cotton stalks contain about 40% of cellulose and 1320% of pentosans, mainly represented by xylan. According to A.S.Sadikov and B.O.Bogamov In addition to pentosans and fibre, stems contain lignin (more than 20.0%), protein (3.0%), amino acids (1.2%), mineral nitrogen (0.3%), monosaccharine (3.5%) and ash (4%). The ash of stems contains such chemical elements as phosphorus, sulphur, chlorine, silicon, calcium, magnesium, potassium, sodium, iron, manganese, boron, copper and zinc [137; p. 1115].

As we can see, cotton stems are very complex in composition and are superior to wood in this respect. However, the main components of cotton stems are the same as those of wood (Table 3.1)[138; 14].

Table 3.1.

Chemical composition of cotton wood and stems

Components, %	Wood (aspen)	Cotton stalks
Cellulose	43,3	up to 40
Lignin	27,5	up to 20
Pentosans	10,4	up to 18
Other connections	18,8	up to 22

Cellulose is the main substance of stems, providing its elasticity and mechanical strength, Its tensile strength in fibre ranges from 2.5 x Yut to 6 x 10^4 N/cm^2 [139; p. 114-120] (260.3 - 624.7 MPa). Cellulose has sufficient resistance to thermal effects. Under certain conditions cellulose is hydrolysed to become monosaccharides. Cellulose molecules comprise 200-3500 homogeneous molecular groups united in a thin (5.7 x 10^7 cm), stretched in one direction, long (1 x $10^{'5}$ -1.8 x $IO^{'4}$ cm) filament [140; p. 50-250].

In the process of hydrolysis, under the action of ultraviolet rays (photochemical degradation), prolonged action of elevated temperatures (thermal degradation) and mechanical effects (mechanical degradation), glucose bonds are broken, as a result of which the macromolecule is broken into short chains [163; p. 200-210].

Lignin is a colloidal substance and under certain conditions acquires the functions of a binder. Cells are bound into a single framework by intercellular substance containing mainly lignin (60-90%) [142; pp. 4-8,   14-16].   Pentosans   enhance   elasticity, flexibility, and

resistance to rupture [143; p. 8-16, 210-230]. When pentosans are hydrolysed, sugar is formed.

Let us consider the structure of cotton stalks pruning. On an end section of a cotton stem, starting from the periphery to the centre, one can distinguish bark, cambium, wood and heartwood. In the upper part of the stem, the woody tissue forms a heart-shaped tube filled with loose tissue of primary formation; in the lower part, the heartwood is more solid and in appearance almost does not differ from the adjacent woody tissue [144; pp. 416-418]. The pith occupies a small part of the stem by volume (5%) and consists of thin-walled parenchyma cells. The woody part contains moisture-conducting channels parallel to the stem axis, which determines good permeability by the binder and optimal heat transfer. Wood

Different rocks are divided into ring-vascular and disseminated-vascular (capillary-porous)   rocks.

increased permeability to liquids and gases.

The chemical composition and physical and mechanical properties of cotton stalks were studied to obtain wood-plastic board materials.

According to the research results of L.T. Ermatov and A. Abdullaev, cellulose provides the stems with elasticity and mechanical strength lignin, which is a colloidal substance, under certain conditions acquires the functions of a binder. Pentosans enhance elasticity, flexibility and resistance to tearing. The chemical composition and

31

ultramicroscopic structure of cotton stems showed that they are closer to hardwood in these two respects.

The acidity (pH) of wood is known to influence the resin curing process and, consequently, the properties of particleboard. The pH of wood of different species varies between 3.3-5.15. The woody part of cottonwood stem has an acidity of 4.8 per cent and the bark has an acidity of 5.1 per cent. This also indicates the proximity of cottonwood properties to those of wood [145;p. 5].

Table 3.2 summarises the results of studies on the isogeny of physicomechanical properties of cotton stalks.

Table 3.2.

Physical and mechanical properties of cotton wood and stems

Physico-mechanical property	Wood (aspen)	Cotton stalks
Volumetric 1 weight in _____ . ' dry state 2/cm	0,38-0,42	0,39-0,47
Tensile strength MPa	0,60-0,68	0,580-0,766

Experimental study has established that the volume weight of cottonwood stems in dry state is 0.38-0.42 g/cm3, and the bending strength is 0.60-0.68 MPa. The specified characteristics for **aspen wood are respectively 0.39-0.47 g/cm3 and 0.580-0.766 MPa. The average moisture content of cotton stalks is 10 %.**

Studies of physical and mechanical properties of cotton stalks are noted in the works of the authors [146; p. 80-86]. According to them, the moisture content of cotton stalks on average is 10%, the bulk weight of different fractions of crushing from 0.224 to 0.196 kg/l, compressibility - 96%, elasticity - 16%. The physical and mechanical properties of cotton stalks such as stiffness, coefficient of friction, moisture content, etc. were also studied. Stiffness and modulus of elasticity of cotton stalks in bending depending on moisture content are given in Table 3.3.

Table 3.3.

Effect of moisture content on stiffness and modulus of elasticity in bending of cotton stalks

Humidity, %	Stiffness, kg/cm^2	Modulus of elasticity in bending, MPa
12	2950,36	9993
34,7	1802,34	7663
43,8	1555,24	6612,4
51,2	1555,24	6612,4
55,8	1506,50	6405,2
61,2	1459,66	6206,04

Table 3.3 shows that with increasing moisture content from 12 to 61.2%, the stiffness and modulus of elasticity of guza-paya decreases by 1.6 times with an average stem

diameter of 1.5-2.5 cm. The volumetric weight of the stems ranges from 69-70 kg/m^3 .
Depending on the season, the humidity of guza-pai ranges from *9 to* 65%.

Fig. 3.2 shows the dependence of elastic modulus of cotton stalks on moisture content
at its elevated values.

At the same time, the elastic modulus of guza-paya increases with decreasing moisture
content due to stem odering.

The deflection and Poisson's ratio as a function of the moisture content of cotton stalks
have also been studied [147; p. 4].

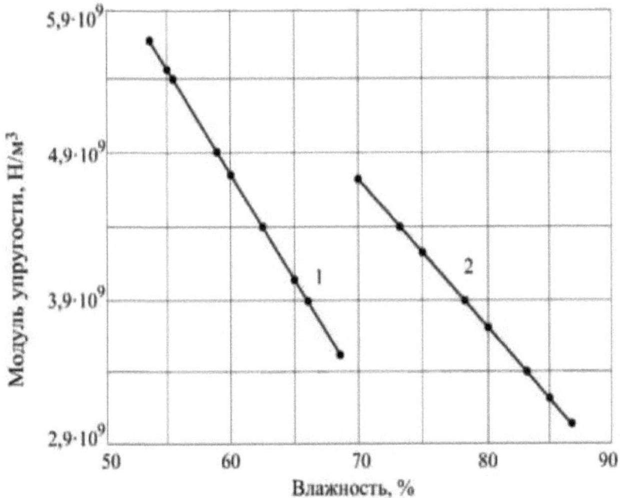

Fig. 3.1. Dependence of elastic modulus of cotton stalks on

dampness

The tensile strength [148; p. 4] of the cross section of cotton stalks depending on its
diameter was also investigated (Table 8)

Table 3.4.

Dependence of tensile strength on the diameter of cotton stalks

Stem diameter, mm	7	9	11	13	25	27
Geometric area	38,5	63,5	94	132	176	226
Force, kg	78,0	60,2	146	338	510	360
Tensile strength, kg/cm^2	2,2	1,0	1,55	2,55	2,9	2,48

The author [149; p. 4] has studied the stiffness of the stem of leafless cotton plants
in bending, determined by vibration method in the post-frost period on grade 18
and at a stem moisture content of 25-30% depending on the diameter is given in Table
9.

Table 3.5.

The dependence of stiffness on the physical diameter of cotton stalk.

Physical	5,5	6,45	6,555	7,35	7,8	8,45	8,7	10,6

diameter, mm								
Y.bRigidity UE, kg/cm^2	339,85	727,12	638,55	721,622	877,92	1882	2135	2630

Analysing the composition, structure and physical and mechanical properties of cotton stalks we can draw the following conclusions:

cotton stalks, especially the woody part, are close in their properties to hardwood and can be used in the production of wood-based panels;

unlike wood waste commonly used in particleboard production, the stalks of valley cottonwood will be shredded together with the bark, as it makes up part of the volume;

In the production of board materials, the size and shape of the particles are subject to stringent requirements. Consequently, not just small pieces but also particles of a certain dispersibility and shape must be obtained from the cotton stalks.

It should also be noted that the existing methods and installations do not provide the necessary chips from cotton stalks and, accordingly, do not allow obtaining conditioned wood-fibre mass for the manufacture of composite wood-plastic board materials with high physical and mechanical properties.

All this requires a new approach to the problem of shredding, and it can be solved on the basis of development of a new method of technology of cotton stalks shredding and obtaining conditioned wood-fibre mass from them.

§ 3.2 Development of a method of grinding cotton stalks, which makes it possible to obtain conditioned wood-fibre pulp from cotton stalks, meeting the requirements of wood-plastic board materials production

As noted in the analysis of the current literature cited in Chapter One, many scientists have been involved in the development and production of particleboard based on cotton stalks. Some scientists have obtained particleboard [150; p. 299] only by using the woody parts of cotton stalks. Others used the woody part of cotton stalks together with wood chips to obtain particleboards. However, in both cases the issue of separating the bark from the woody part was not solved. In addition, the bark and small parts of cotton stalks remained as production waste [150; p. 300].

In recent years, studies conducted by scientists in the field of cotton stalk shredding on existing serial equipment have shown that none of the existing types of shredders can produce conditioned wood-fibre mass from cotton stalks without waste, as they differ significantly from wood and other wood waste by the presence of strong and elastic stem bark. The fibres in the bark envelop the cutting tools of the shredder, which leads to frequent machine stoppages or breakage of the cutting tools and their replacement.

Thus, analysing the composition, structure and physical and mechanical properties of cotton stalks we can note the following:

cotton stalks, especially the woody part, are close in their properties to hardwood and can be successfully used in the production of wood-based panels;

Unlike wood waste, which is commonly used in the production of particleboard, cotton stalks must be shredded together with the bark, as it makes up a significant part of the volume;

in the production of wood-plastic board materials, stringent requirements are placed on particle size and shape [151;p. 4].

Consequently, not just small pieces, but particles and fibres of a certain dispersibility and shape, i.e. conditioned wood-fibre mass, must be obtained from cotton stalks.

All this requires a new approach to the problem of grinding cotton stalks and it can be solved by developing a new and effective method of grinding and obtaining conditioned wood-fibre pulp, allowing to obtain composite wood-plastic board materials with high physical and mechanical properties [152; p. 68-70].

Taking into account the above theoretical and practical analyses, we have developed a method of two-stage grinding of cotton stalks, which allows us to obtain conditioned wood-fibre mass by cutting without any waste.

In the first stage, in order to obtain a conditioned chip, the cotton stalks were cut together with the fibrous part (bark) of the part at a certain length with an equal cut without end edges. In the second stage grinding of wood chips was carried out and a conditioned wood-fibre mass was obtained, allowing to obtain composite wood-plastic board materials.

Figure 3.1 shows a schematic representation of a modernised plant for shredding cotton stalks and obtaining chips from them. It includes the following assemblies and parts: table 1, frame 2, disc saw 3 with a drive (not shown in the figure) mounted on table 1, fixed on frame 2, as well as guide rail 4, which has the possibility of horizontal movement in order to adjust the length of chips when cutting cut wood, top bar 5, mounted above the cutting area and hopper 6 for chips.

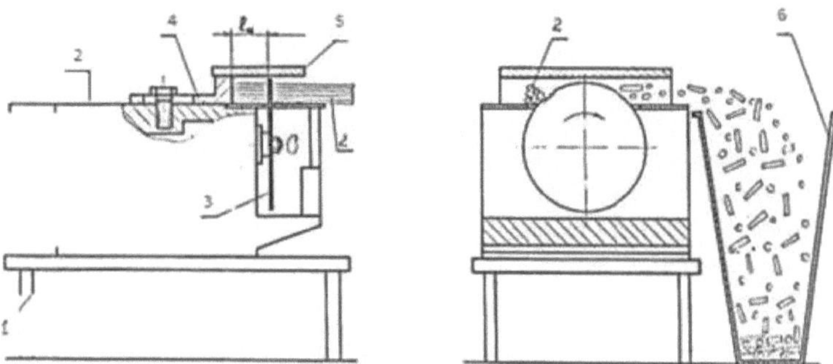

1 - table; 2 - frame; 3 - circular saw; 4 - guide rail; 5 - bar; 6 - **chip** *hopper*
Fig. 3.1. Schematic representation of the installation for chip production from cut wood

The operation of the plant is as follows:

The stem bundle 2 is fed into the cutting zone and passes through the protruding sector

The saw is mounted under the machine table. The saw is set at a certain height above the table surface.

The gap between the saw and the guide bar gives the chip length (1sh). When shredding cotton stalks, the resulting chips will have a directed movement along the vector of linear speed of the saw, which is ensured by the upper bar. At the end of the table there is a hopper 6 for chip collection.

Based on the developed drawings, the plant was manufactured in the special design office of KV-KOMRO21T Ltd.

Figure 3.2 shows a schematic representation of the upgraded plant for chipping chips by cutting and obtaining wood-fibre pulp for wood-plastic composite board materials.

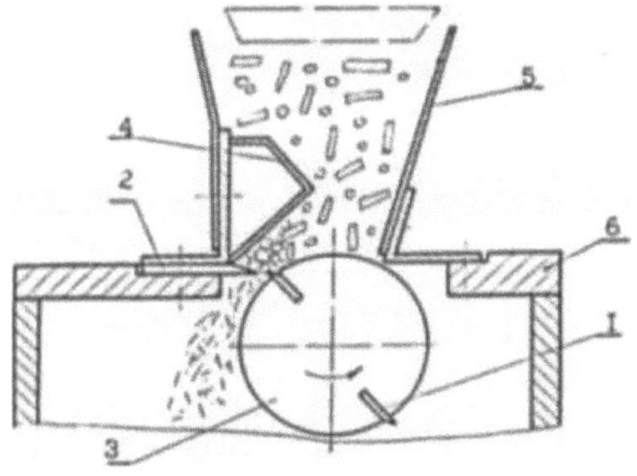

1-blade; 2-counter-blade; 3-knife drum; 4-pads; 5-hopper; 6-machine table;
7-hopper accumulator

Fig. 3.2. Schematic representation of the installation for obtaining woody

fibre pulp from cotton stalk chips

Schematically, the device is shown in the figure and includes the following assemblies and parts: knife 1, counter-knife 2, knife drum 3, lining 4, hopper 5, machine table 6.
The principle of operation of the plant is as follows:
The cotton stalk chips are fed into hopper 5 and then into the cutting zone
between the drum knife and the counter-knife, and then as wood-fibre pulp into the storage hopper.
The specified modernised units on the basis of the developed working and construction drawings were manufactured at the special design bureau of LLC "KV-KOMRO21T" and tested in laboratory conditions of cotton stalks grinding and wood-fibre mass fractional composition and distinctive characteristics of which are shown in Figure 3.3. Distinctive characteristics of crushed cotton stalks, such as the presence of fibrous part, heartwood and heterogeneity of size and shape of wood particles are considered (Figure 3.3).

**Figure 3.3. Fractional composition of chopped stem mass
cotton**

Note: in Figure 3.3, the numbers in the corner of the photograph mean the following: the first number is the size of the sieve mesh in millimetres through which the given fraction has passed; the second number is the size of the sieve mesh in millimetres on which the given fraction is selected; the third number is the length of the chips in millimetres from which the given chip is obtained.

Figure 3.3 shows that the crushing of cotton stalks produces a wood-fibre mass consisting of needle-shaped wood particles, fibrous inclusions formed from the bark, and a fine fraction consisting of crushed wood and stem core.

Each of these components has its own strength properties, physical characteristics and chemical composition, particle size and shape. Thus, the study found that the bulk weight of different crushed fractions was from 0.224 to 1.96 kg/l, compressibility - 90%, elasticity -1.6%. This circumstance is the main distinguishing feature of filler from cotton stalks and requires studying and adjusting all technological modes of production of board material.

Complex in composition and heterogeneous in particle size and shape, especially due to the presence of fluffy bast fibre, the filler behaves differently than wood chips during mechanical and pneumatic separation and transport, which also required a special study of the bulk density, volatility and fractionation of shredded cotton stalks.

The study of pneumatic separation of chip mass from cotton stalks in laboratory conditions has shown that at certain particle dispersity the constituent parts of the stalk, such as wood chips, free fibre and dust particles, are distributed in the blowing zone sequentially - the smallest particles are located in the most remote area from the place of mass feeding, then a zone consisting of fibre is formed and wood particles fall closer to the centre.

Known installations for carpet formation work on the principle of scattering the mass under the air jet in two directions - in the course of the pallet movement and against it. In this case, a multilayer carpet is formed from different sizes of wood particles. In our case, a carpet is formed with the largest wood particles in the centre, followed by successive layers of smaller wood particles and fibres and dust in the outer layers. A thin layer of the finest particles gives the board a smooth surface. The fibres forming the main outer layer, due to their high tensile strength, give the board an increased bending strength, which is 20-25 % higher compared to a single-layer board.

§ 3.3 Conclusions to the third chapter

1.   The compositions of physicochemical and strength properties of cotton stalks have been studied. It was found that cotton stems consist of 50% stem wood, 6% bolls, 30% bark and 14% root part. It also consists of 40% cellulose, 13-20% pentosan, 20% lignin, 3.0% protein, 1.2% amino acid, 0.3% mineral nitrogen, and 4% ash. Cotton stalk ash also contains elements such as phosphorus, sulphur, chlorine, silicon, calcium, magnesium, potassium, sodium, iron, manganese, boron, copper, zinc.

2.   This paper presents a two-stage method and plant for grinding cotton stalks to produce conditioned wood fibre pulp for use in the production of composite wood-plastic board materials.

3.   It was also found that the volume weight of cotton stalks in dry state is 0.38-0.42 g/cm and the bending strength is 0.60-0.68 MPa.

4.   It is shown that cotton stalks depending on its moisture content from 12 to 61,5%, stiffness is from 2950,30 to1459,66 kg/cm^2 , and modulus of elasticity in bending (t = 7mm) from 9993 to 6206,04 MPa. The bulk weight of different chipping fractions is from 0.224 to 0.196 kg/l, chip compactability is 96% and elasticity is 16%.

5.   A two-stage method and installation for grinding cotton stalks and obtaining wood-fibre pulp for use in the production of composite wood-plastic board materials have been developed.

CHAPTER IV

STUDY OF THE PROCESS OF GRINDING
COTTON STALKS
AND OBTAINING A CONDITIONED
WOOD-FIBRE MASS OF FILLER ON THEIR BASIS FOR
USE IN THE PRODUCTION OF WOOD-PLASTIC
COMPOSITE BOARD MATERIALS.

Further, the results of research into the influence of the main technological factors [160; p. 14] of the shredder design on the process of shredding and formation of conditioned wood-fibre mass on the basis of the developed two-stage shredding are considered. In the primary shredding of cotton stalks, wood chips are obtained and in the secondary shredding of wood chips from cotton stalks, wood-fibre mass for the production of wood-plastic board materials is obtained [153; p. 358-359].

First of all, it was necessary to determine the main criteria for the conditioned wood chips, i.e. their suitability for obtaining quality wood fibre pulp at repeated milling. It was theoretically substantiated that in order to obtain conditioned wood-fibre pulp the chips should be of a certain length, have no "tails" of bast fibre at the ends, and contain no separated long fibre in the mass. This requirement is achieved by using a multi-saw cutting device with a set distance between the saws and quality sawing, which gives an even cut at the ends of the stem section and excludes stripping of bark from the stem.

Sawing is a process of closed cutting with multi-blade tools-saws that separate wood into volume undeformed parts by turning the nominal volume of stems between these parts (kerf) into chips. In crosscut sawing, the lateral [154; pp. 29-30] blades of the tooth cut the fibre and form the walls of the kerf, while the front surface shears the cut fibres and forms the bottom of the kerf. This defines the following requirements for [155; p. 5-8] tooth geometry: the side blade must cut the fibre before the front surface makes contact with it.

For this purpose, it must be extended forward along the course of the saw relative to the short blade due to the negative (or zero) contour angle and have a positive angle due to the oblique sharpening [156; p. 128]. The deeper the indentation of the tooth, the greater the work of friction, and hence the greater the cutting power consumption. The coefficient of friction of cotton stalks on metal is in the range of 0.32-0.53 depending on the moisture content of the stalk and the type of metal. Compared to wood, however, it is greater, so tooth penetration into the stalk should be limited compared to wood. Since the stalks have on the surface a strong elastic bark and under it a woody and loose part, and also the bundle contains many stalks not connected to each other, rigidly, the mass of cotton stalks more resembles an elastic-viscous body than a solid one, creating a stress state through deformation [157; p. 126].

At low cutting speeds and insufficient feed rates, viscous deformations have time to develop on the tooth. The resistance of an elastically viscous material to external action is related to the speed of stress and strain propagation in it. Therefore, quality sawing of the stalk and obtaining of conditioned chips mainly depends on the correctly selected feed and cutting speed, moisture and density of the cotton stalk bundle [158; p. 86-87].

In this regard, the influence of cutting speed, moisture content and degree of density of the stem bundle on the quality of conditioned wood chips obtained from it was further investigated.

<h3 style="text-align:center">§ 4.1   Investigation of the process of primary shredding of
cotton stalks
and obtaining chips from them</h3>

<h3 style="text-align:center">§4.1.1 Investigation of cutting speed on the quality of produced chips
from cotton stalks</h3>

Cutting speed. The speed of material deformation during cutting influences the quality of the obtained surface. As the cutting speed increases, the inertia force of the stem mass in the layer below the cutting surface increases, which reduces the deformation in it, creating a "back-up".

The stresses arising at the point of contact between the blade and the layer are transmitted with some velocity into the layer. The velocities of stress propagation correspond to the velocities of strain propagation as well [159; p. 463-464].

In elasto-viscous material the velocity of stress propagation is low, so the impact of the saw tooth is transmitted to the layer slowly and, consequently, at a higher speed of this impact the stresses arising from it are concentrated and localised in the place of impact, which causes local fractures with less energy consumption [160; p. 8992].

It has been proved that an increase in cutting speed from 40-50 m/s to 100 m/s can cause a 30-40% increase in cutting force.

Localisation and concentration of destructive energy at the saw edge, reduction of material pre-compression by the disc, increase of inertial support, reduction of friction coefficient - all these consequences of cutting speed increase have an external manifestation, which is of the most importance for this research. This is an increase in the cleanliness of the cut, in which the cut surface is reduced due to the reduction of macro-overhangs and depressions [161; p. 5-8], reduction in the number of deformed particles, "shanks" on the chip ends, bark fibres and long particles - which is ultimately defined as "chip conditionality".

During the experimental studies, the cutting speed was varied in the range of 15-50m/s. Since cutting conditions vary depending on such factors as stem moisture and degree of compaction, the experiments were conducted at different values of these values.

At the same time, the mutual influence of feed rate and cutting speed, expressed through the quality of chipping, i.e. the content of conditioned particles in chips B (%), was studied.

Thus, when determining the effect of cutting speed on B, the feed rate varied from 0.02 m/s to 0.2 m/s.

Fig. 4.1a shows the dependence of B on cutting speed at different feed rates U and 10% moisture content of stems, Fig. 4.1.6 shows the dependence of B on cutting speed

41

at 60% stalk moisture content.

It is evident from the fig. that at all feed rates with increasing cutting speed the amount of conditioned chips in the mass increases, i.e. the cutting process improves. Curves have parabolic character The greatest amount of conditioned chips is at y=0,02*0,12 m/s, further increase of y leads to decrease of B. It testifies to violation of cutting mode due to deep penetration of saw teeth into the layer of stems. Characteristically, the influence of cutting speed on B at low feed rates is more pronounced, which indicates the predominant influence of cutting speed on the material at feed rates that can be neglected.

Since increasing the cutting speed without limitation leads to energy costs and wear of the cutting unit, it is reasonable to choose the speed limits that correspond to the highest. This optimum zone in Fig. 4.1a corresponds to the value of B - 75-80%.

When changing the stem moisture content W, the character of curves changes insignificantly (Fig. 4.1b). Only quantitative values of the studied factors change. It can be concluded that in the investigated W interval it is possible to obtain conditioned chips. The influence of humidity on the quality of chipping is discussed in detail below.

On the basis of this part of the work it can be concluded that for cotton stalks cutting speed and feed rate have a limited range of values, beyond which the quality of the crushed mass and the process as a whole decreases. This is due to the natural properties of cotton stalks and the need to pulverise them in a mass (bundle).

Thus, the established optimum cutting speed is 32.5 m/s feed rate is 0.04-0.16 m/s.

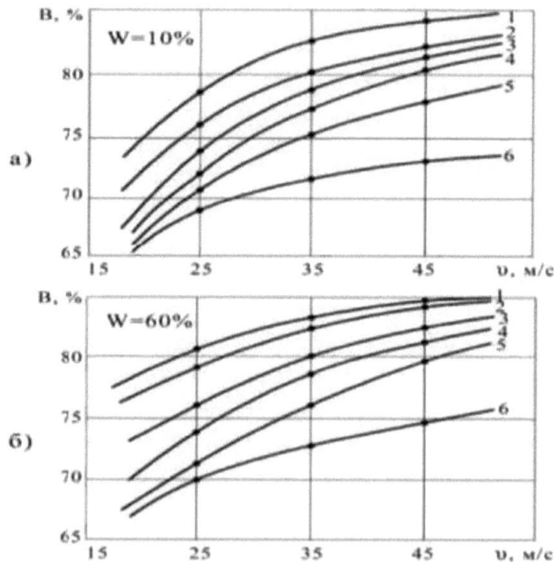

Fig. 4.1. Dependence of the amount of conditioned chips on the cutting speed at feed rates in m/s: 0,02 (1); 0,04(2); 0,08(3); 0,12(4); 0,16(5); 0,2 (6)

Studies of the influence of feed rate on the productivity of the process and cutting power showed that at the maximum value of feed rate (0.16 m/s) in the selected interval productivity is equal to 50kg/h, power consumption 0.7 kW on a laboratory unit with a double core drum.

§4.1.2 Investigation of the influence of cotton stalks moisture content on the quality of produced chips

Stem moisture. The moisture content of the stalks has a great influence on the sawing process, as the mechanical properties of the stalks depend on it to a large extent.

Stem resistance against mechanical impact is determined not only by physical and mechanical properties of tissues, but also by the hydrostatic pressure of free moisture in the stem cell cavities.

The authors studied the stiffness and modulus of elasticity of stems from moisture content. It was obtained that both indices decrease by 37% when humidity increases from 12 to 60%.

Moisture also affects the coefficient of friction of the stalks on the steel, which in turn affects the quality of grinding. The Poisson's ratio increases with increasing moisture content. Obviously, the influence of stem moisture on the process of chip formation, as the shape, size and fractional composition of the chip mass change according to the moisture content. All this indicates the need to study the influence of cotton stalks moisture content on the shredding process [162; p. 207-208].

Fig. 4.2 shows the dependence of conditioned chips - B on stem moisture at different feed and cutting speeds. It can be seen that the increase in moisture content up to 40% leads to an increase in the conditioned wood chips - B then, passing through the maximum, the curve smoothly decreases. The optimum zone for 73 conditioned wood chips - B falls on W =30-50%. At the same time, the process is most efficient at the feed rate of 0.04-0.12 m/s.

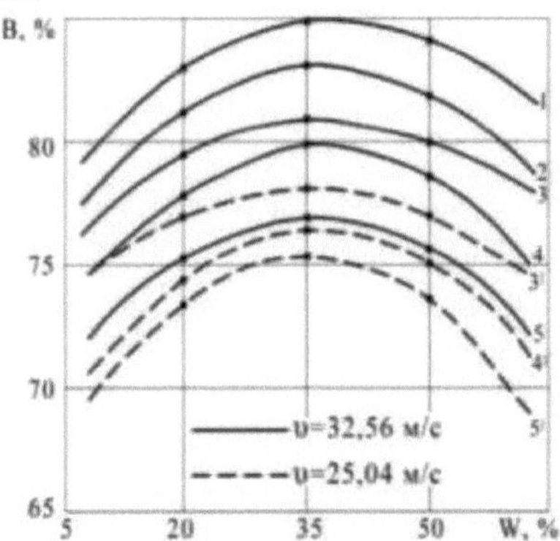

Figure 4.2. Effect of cotton stalk moisture content on the number of

of conditioned wood chips at feed speeds: 0.04 m/s (1), 0.08 m/s (2), 0.12 m/s (3.3) 0.16 m/s (4.4) 0.2 m/s (5.5)

From the graph we can conclude that the lower the moisture content, the higher the speed should be in order to meet the requirement of B=75-80. So at moisture content of 10% it is achieved at feed speeds of 0.040.12 m/s. The same picture is observed at moisture content above 50%. Since the moisture content of the stems varies from 10 to 60% depending on the period of harvesting and storage, it can be concluded that a special drying or humidification process can be avoided. Similar results are obtained for secondary chipping.

§4.1.3 Investigation of the influence of the degree of stem bundle density on the quality of obtained chips

The degree of density of the stem bundle, expressed through the compaction factor), is of decisive importance in closed chopping 74

by cutting. As j increases, the cutting speed increases. In a sufficiently compacted layer, direct support is formed by the layers of stems against each other. In a loose layer with a low compaction factor j, the "inertial support" becomes more important. That is why in this type of layer, the increase in cutting speed is accompanied by a significant drop in the value of pre-compression and, accordingly, an increase in saw operation [163; p. 428].

In a sufficiently compacted package, the stems are rigidly fixed in interaction with the working tool and there is no vibration of the stem, which ensures high-quality cutting and prevents the formation of long pieces of stem and bark.

To determine the optimum beam density, the experimental setup is equipped with special pushers between which the stems are placed parallel to each other and perpendicular to the saws, the fixation from above is carried out by   adjustable

The artificial chamber is formed with an area of the side wall Sh =lh. An artificial chamber is formed with side wall area S_H =lh

In general, the compaction factor is characterised by the ratio of the total cross-sectional area of stems to S_H and is expressed by the formula:

$$j = n\ d_{cp}^2\ i_{cp}/4lh$$

Where d p_C^2 -- the average diameter of the stems,

i_{cP} -- average number of stems in the chamber.

l,h -- length and height of the side wall of the chamber.

By its value] can be taken as the beam density.

Reznik N.E. found that during beam compression the compaction factor cannot be higher than 0.7-0.8. In this study the amount of conditioned wood chips B was chosen as a criterion of process efficiency.

The obtained results showed that the dependence of j on c is linear in the investigated range of j values. Therefore, when determining the optimal compaction factor, we chose the smallest value that satisfies the conditions for obtaining B=75-80%. Such value was j=0,5-0,65.

The work required to reach the set point can be taken as the feed force when calculating the feed kinematics of the device.

Thus, in the first part of the work, theoretical and experimental analyses of the process of primary grinding of cotton stalks were carried out, which allowed us to determine the main factors affecting the efficiency of grinding. The optimal parameters of such factors as stalk moisture, bundle density, cutting speed and feed rate were determined. On the basis of the conducted researches the technology of primary shredding based on sawing of a bundle of cotton stalks by a multi-saw working organ is developed, the regularities of the shredding process for obtaining chips suitable for further processing in the production of composite boards or other materials or products are established. The laboratory and semi-industrial installation for primary shredding of cotton stalks protected by the copyright certificate [164; p. 229-130] is created.

§ 4.2   Analysis of the obtained results of research of the first and second grinding process

and development of a modular line for grinding

cotton

stalks in

order to obtain conditioned shredded wood pulp

and fillers from them for the production of composite

wood-plastic board materials.

Based on the study of numerous literature sources in the field of theoretical and practical shredding of wood and stems of annual plants, including cotton stems, a requirement for obtaining chips from them was developed. The chips should be of a certain length, have no bast fibre "tails" at the ends, and contain no separated long fibre in the mass. This requirement is achieved through the use of multi-saw cutting organ with a set distance between the saws and quality sawing by cutting, which gives an even cut on the ends of the cotton stalks cut and accordingly excludes stripping of bark from the cotton stalk [166; p. 129-133].

It has been shown that obtaining of quality conditioned chips during primary chopping of cotton stalks depends largely on the correct choice of feed and cutting speed, as well as on the moisture and density of the cotton stalk bundle. Experimentally it has been established that quality conditioned chips can be obtained at cutting speed - 32,5 m/s and feeding speed of cotton stalks - 0,04-0,16 m/s depending on moisture and density of cotton stalks bundle. The optimum content of conditioned chips is observed at moisture content of cotton stalks within $W = 30$-50%. Thus the most effective process goes at speed of feeding 0,04-0,12m/s. As humidity of cotton stalks depending on speed of harvesting and storage varies from 10 to 60%, it is possible to draw a conclusion that special process of drying of cotton stalks or moistening can be avoided. At the same time, it was found that the optimal degree of bundle density is] = 0.5-0.65.

On the basis of results of theoretical and practical researches the installation for secondary shredding of cotton stalk chips with a special overlay on counter-blades is developed and created. At the same time it is determined that the angle of overlapping of the unit should be $p = 30$-$50°$ for different chip moisture content. The time consumption will be the least.

It is also established that the best conditioned wood-fibre mass - filler at secondary grinding of stems is obtained at chip length of 25-40 mm, allowing to obtain composite wood-fibre board materials with high physical and mechanical properties.

It should also be noted that we consider it necessary to dwell in more detail on the results of research on the influence of cutting speed on the quality of the obtained chips from cotton stalks. It is revealed that the increase in cutting speed more than 40-50 m/s can cause an increase in cutting force by 3040%. Therefore, the experiments of research into the influence of speed on cutting of cotton stalks for obtaining conditioned chips from it were carried out within the interval 15-50 m/s.

It has been established that the optimum cutting speed of cotton stalks for obtaining chips from it is 32.5 m/s, and the feeding speed of a bundle of cotton stalks is 0.04-0.16 m/s. It should be noted that at the maximum value of feed speed - 0.16 m/s and the selected interval productivity is equal to 50 kg/h, power consumption 0.7 kW on the created laboratory installation with two-core drums.

The influence of cotton stalks moisture on the process of chopping and chip production at natural moisture of stalks up to 75-80% has been studied. At the same time it is established that at grinding of cotton stalks a special process of drying or moistening can be avoided. Similar results are obtained for wood-fibre pulp at secondary shredding.

The compaction factor is characterised by the ratio of the total cross-sectional area of the stalks at the length and height of the side wall. The compaction factor of cotton stalks at a moisture content of 75-80% is 0.5-0.65.

The second pulverisation of wood chips was carried out on the created plant, the principles of which are shown in Figure 4.3 [167; p. 129-133].

It was found that the length of the chips has an influence on the length of the woody part of the stems and on the quality of the obtained wood fibre pulp. According to the requirement the length of the received chips should be within 4-6 cm, as further increase leads to obtaining of uneven cutting of stems and obtaining of not conditioned wood-fibre mass and accordingly obtaining of composite wood-plastic board materials of low quality. The study found that the best composition of wood-fibre mass were obtained at the length of chips 25-40 mm.

It is also revealed that quality wood fibre from cotton stalks can be obtained when the value of the gap between the knife and counterbody is within 1.2 mm.

§ 4.3 Conclusions to chapter four

In this fourth chapter, the results of primary and secondary shredding of cotton stalks and obtaining of conditioned wood fibre pulp - filler for the production of wood-plastic board materials are discussed.

At primary shredding of cotton stalks at moisture content of 70-80% the optimum length of chips from cotton stalks is 25-40 mm, the optimum cutting speed of cotton stalks is 32.5 m/s, and the speed of cotton bundle feeding for cutting is 0.04-0.16m/s. At this mode of cutting cotton stalks, a conditioned chip is obtained from it.

High-quality wood-fibre pulp was obtained at chip length at the established unit of secondary chipping of cotton stalk chips at the value of the gap between the knife and counter-body within 1.2 mm at chip lengths in the range of 25-40 mm.

Thus, on the basis of theoretical and practical studies of the process of cutting wood and stems of annual plants, as well as the developed new method of crushing cotton stems, an upgraded unit for primary crushing of cotton stems and secondary crushing of chips from the stems has been created, which allows to obtain conditioned wood-fibre mass for production and application in the production of wood-plastic composite board materials.

CHAPTER V

PRACTICAL AND ECONOMIC ASPECTS OF THE DEVELOPED TECHNOLOGY OF COTTON STALKS GRINDING AND OBTAINING ON THEIR BASIS WOOD-FIBRE FILLER FOR THE PRODUCTION OF COMPOSITION WOOD-PLASTIC PLASTIC BOARD MATERIALS

§ 5.1 Technology of obtaining composite wood-plastic board materials on the basis of wood-fibre mass-filler obtained by the developed two-stage method of cotton stalks grinding.

On the basis of the above patent-licence study and literature analyses of numerous data we have developed scientific-methodological and technological principles of obtaining wood-plastic composite board materials based on wood fillers from cotton stalks and polymer binders [168;p. 10-11].

Figure 5.1 shows the technological scheme for obtaining wood-plastic composite board materials [169; p. 5-6].

The principal technological scheme of wood-plastic composite board materials production based on the developed scientific and methodological principles includes the following main stages of the technological process:

primary crushing of cotton stalks into wood chips;

secondary grinding of wood chips into a conditioned wood-fibre mass - filler;

fractionation of wood-fibre pulp into fractions; processing (introduction of binder components) of chips in a container with a doser; drying of wood-fibre pulp from cotton stalks; mixing of wood-fibre pulp with polymer binders in a DCM-type mixer.

Fig. 5.1. Principal technological scheme of wood-plastic composite board materials production on the basis of wood fibre mass - filler from cotton stalks and polymer binder

Formation of osmolised wood fibre pulp to produce

wood-plastic composite boards;

pressing of wood-fibre pulp and production of wood-plastic board materials;

cooling of compressed wood-plastic board materials; storage of finished wood-plastic board materials products.

Based on these scientific and methodological principles, we have created an improved technological line for the production of wood-plastic composite board materials based on wood fillers from cotton stalks and polymer binders [174; p. 29-30]. It makes it possible to obtain wood-plastic board materials and structures based on them with high physical and mechanical properties that meet modern requirements. On the basis of the conducted research we can draw the following conclusions On the basis of the above patent and licence study and literature analyses of numerous data we have developed scientific, methodological and technological principles of obtaining wood-plastic composite board materials based on wood fillers from cotton stalks and polymer binders.

The technological properties of chip mass have been investigated and the optimum technological modes (pressure and time) of composite pressing at obtaining wood-plastic composite boards have been determined.

The obtained general tendency of change of physical and mechanical properties of the board at different densities and binder contents gives an opportunity in practice to rationally select the required density and binder content depending on the requirements to the boards and their purpose.

The influence of chip moisture content (before grinding and after mixing with binder) on the quality and properties of board materials has been studied. The moisture content should be in the range of 9-10%.

It is established that the density of the material should be in the range of 700-750 kg/m^3 , with a binder content of at least 10 % should be 700-850 kg/m^3 . Acceptable values of particle sizes and their influence on physical and mechanical properties of board material have been determined.

In order to implement the production technology of composite wood-plastic board materials, first of all, in addition to the technology of obtaining wood-fibre mass-filler from cotton stalks, it is necessary to determine the technological modes of pressing osmolonised wood-fibre mass. Since the pressing process has a significant impact on the formation of such qualitative indicators as: bending strength, tensile strength at rupture perpendicular to the plate, density, hardness, modulus of elasticity, specific resistance to pulling nails and screws, water absorption and swelling 170; pp. 9-12].

In connection with the above-mentioned, during the development of the technology for obtaining plates, special attention was paid to the study of pressing modes to establish the optimum value of the specific pressure during pressing of plates and the duration of pressing time, as well as the temperature of pressing.

In accordance with the scientific and methodological principles of the laboratory, the images of wood-plastic board materials were obtained experimentally. At the same time pressing and its modes were carried out according to the recommendation of D.K. Kholmuradova on hydraulic press P-250 with electric heating. We have established that the value of specific pressure should be within 3.0 and 3.5 kg/cm^2 , at the temperature of plate heating within 170 - 18oC. The optimum time of staying of the press composition 84 under the press pressure (3,5 kg/cm^2) is 10 min.

Thus, on the basis of the conducted complex works the optimal mode of pressing of wood-plastic composite board materials was determined, which consists of the following: specific pressure of pressing - 3,0 kg/cm^2 ; pressing temperature 170° C; duration of pressing with heating - 10 min.

Table 5.1 shows the comparative data of physical and mechanical properties of particleboard and the obtained wood-plastic composite board materials with different board densities.

Table 5.1.

Physical and mechanical properties of particleboard and composite wood-plastic boards from cotton stalks and polymer binders at different densities

Material property values	Properties of		

	chipboard, according to GOST 10632 77 at flesh. 720-800 kg/m^3	CP properties at densities, kg/m		
		600	675	760
Bending strength, MPa for 16 mm thickness not less than 16 mm	15-18	17/22	23/28	27/31
Tensile strength perpendicular to the plate formation, MPa, not less than	0,3-0,35	0,45/0,6	0,85/0,96	0,9/1,4
Swelling, % max. at normal water resistance	20-30	27/30	18/27	15/19
Hardness, MPa (approximate)	19,6-39,2	30/35	35/44	38/49
Modulus of elasticity in static bending, MPa	1770-4410	1500/2000	2200/3200	3000/4700
Specific resistance to nail pulling, MPa	2,45-2,65	2,3/2,5	2,53,8	2,6/3,9
Specific resistance to pulling out screws N/m	58800 117700	60000/ 92000	90000/ 115000	100000/ 126000

Note: in numerator: slabs obtained by traditional method; in denominator: slabs obtained by the proposed technology

In this case, the composite board material was obtained in the following

technological mode: pressing temperature 18ooC; pressing time 7 min, pressing pressure 3-3,5 MPa.

As can be seen from the data in Table 5.1, the obtained boards by all indicators of physical and mechanical properties are significantly better than particleboard boards manufactured at the plant EZSP, VNIIDrev, as well as according to GOST 10632-89. Thus, we can conclude that the wood-plastic composite board materials obtained by us according to the optimal technological modes exceed the requirements of GOST 10634-78 and 10637-78 by all indicators.

§ 5.2 Mastering the technology of obtaining composite wood-plastic board materials on the basis of wood-fibre mass - filler from cotton stalks and polymeric binders by organising the production of pilot batches and conducting pilot tests in production conditions.

As noted above, there are a variety of technological equipment for obtaining wood-plastic boards. However, it is impossible to produce high-quality wood-plastic composite board materials based on cotton stalks using the existing technology. In this connection, pilot tests of the technological line were carried out according to the scheme of the technological line developed by us [179].

It should be noted that all stages of the technological process worked satisfactorily. Cotton stalks, in the form of a bundle, were fed to the shredder "Composite-1", developed in the State Unitary Enterprise "Fan va Tarakkiyot". The shredded mass had medium friability, the length of chips was from 10 to 50 mm, the amount of free fibre was more than 22%, the length of which was 20:100 mm, dust-like particles about 19% .

During secondary chipping on the DS-7 machine it was found that due to the presence of fibrous inclusions, the feeding of the material to the auger was accompanied by a slight return of the mass by the scraper and return conveyor to the hopper, which is allowed and provided for by the production technology. The dustiness of the secondary grinding area during the operation of the machine showed the need to install exhaust ventilation.

The particle mass after the upgraded chopping machine DS-7 had satisfactory bulkiness, the fractions of wood and fibre parts practically did not differ in geometric shapes and sizes.

When transported to the drying chamber, the mass was evenly coated and no burning in the drying chamber was observed due to the correctly set drying regime. After sintering of the chip mass, clogging and enveloping of the mixer blades by bast fibres was not observed. The laboratory analysis showed uniform sintering of particles, no lump formation was observed.

Osmolised chips were fed to the moulding machine by means of a belt conveyor, and there was no violation of the feeding mode. The moulding machine operating on the principle of pneumatic separation worked satisfactorily. The carpet was poured in an even, uniform layer. At chip moisture content of 11% there were no blockages and mass hang-ups in the machine units. When the moisture content was increased up to 14%, clogging of the needle roller and registers was observed. That is why the humidity of the osmolised chips, according to the regulations, was kept within the range from 8.8 to 11%. Experimental-industrial tests were carried out on the basis of the developed scientific and methodological technological principles on the technological line 87 for the production of wood-plastic composite board materials of the production site of "PROSPER ALL" Ltd. The chipboard carpet was fed into a hydraulic press. The obtained boards had smooth even surface at visual observation, they differed from particleboards only by dark brown colouring.

In accordance with the developed by us optimal technological mode of production of chipboards from cotton stalks 1008 m^2 wood-plastic composite boards were obtained. Testing of the boards for physical and mechanical properties was carried out in the production conditions of PROSPER ALL Ltd.

Thus, a pilot batch of composite wood-plastic boards under production conditions was obtained.

The manufactured wood-plastic composite board materials had the following average physical and mechanical characteristics:

bending   strength22   MPa,

tensile strength 0.65 MPa,

density57   .0-95. 0kg/cm^3 ,

water absorption37%   ,

swelling27   %.

Table 5.2 shows comparative data of physical and mechanical properties of wood-plastic composite board materials obtained by us in EUP "Fan va tarakkiyot", at the plant EZSP (Tashkent) and VNIIDrev.

As can be seen from the data in Table 5.2, the obtained boards by all indicators of physical and mechanical properties are significantly better than particleboard (particleboard) manufactured at the EZSP plant.

Our production tests have shown the possibility of using the developed technology for the production of wood-plastic composite boards from cotton stalks and polymer binders on the modernised technological lines of "PROSPER ALL" Ltd.

Table 5.2.

Comparative table of physical and mechanical properties of the boards obtained by the pressing regime of GUP "Fan va tarakkiyot", EZSP and <u>VNIIDrev</u>

Physical and mechanical properties of the boards	Pressing mode GUP "Фанва " tarakkiet": P=35 кт/м^2 E=0,31 min/mm T =17O° C	EZSP pressing mode: P=35 kg/m^2 t=0.37 min/mm T = 18O° C	Pressing mode according to VNIIDrev technological instruction P=1,96+2,16MPat =0,3 min/mm T = 170^0 C
Static bending strength, $o_и$, MPa	17,0-27,0	10,1-17,6	14,7-25,5
Tensile strength perpendicular to the slab, Cp, MPa	0,4-0,9	0,2-0,44	0,29-0,39

Swelling in 24 hours AS %	16-30	24-34	20-30
Water absorption in 24 hours, w, %	40-68	66-109	Not regulated

Thus, we can conclude that the developed wood-plastic composite board materials exceed the requirements of GOST 10634, GOST 10637 by all indicators.

It is concluded that in order to obtain wood-plastic composite board materials of good quality it is necessary to store cotton stalks not more than 1 year in dry rooms. Their storage in the open air leads to a decrease in technological properties, i.e. it affects the quality of boards. To reduce dustiness in the chopping zone it is necessary to clean the stalks from sticking earth, to install additional ventilation. To eliminate breakages and new formation in machines it is necessary to provide measures excluding foreign objects in bales.

It can be stated that the obtained results of physical and mechanical properties of wood-plastic boards from cotton stalks and polymer binders indicate the possibility of using the developed technology in the industrial production of wood-plastic boards.

§ 5.3 Development of an enterprise standard (technical specifications) and technological regulations for the production of wood-plastic composite board materials

We have developed technological regulations for the production of wood-plastic composite board materials, which consists of the following points:

1. Characterisation of wood-plastic composite board materials;
2. Technological scheme of production of wood-plastic composite board materials;
3. Technology of obtaining wood-plastic composite board materials;
4. Characteristics of raw materials used in obtaining wood-plastic composite board materials;
5. Control and management of technological process of wood-plastic composite board materials production;
6. Safety technology, fire safety;
7. Environmental Protection;
8. List of production instructions.

Scientific and experimental data testify to the fact that when implementing in practice the technology of obtaining wood-plastic composite board materials, the issue of providing the Republic of Uzbekistan with domestic wood-plastic composite board materials is solved. It should be noted that the present technological regulations include compositions and technology of obtaining, developed by us, necessary for the production of wood-plastic composite board materials [171;p. 158].

The technology of obtaining wood-plastic composite boards from cotton stalks was tested on the production-technological line of PROSPER ALL Ltd. In accordance with the developed "Regulations for the technological process of obtaining wood-plastic composite boards from cotton stalks" 1008 m^2 boards were obtained.

It should be noted that during secondary grinding of wood chips on the modernised machine it was found that due to the presence of fibrous inclusions the feeding of the material to the auger was accompanied by a slight return of the mass by the scraper and return conveyor to the hopper, which is allowed and provided by the technology. The dustiness of the secondary grinding area during the operation of the machine showed the need for ventilation.

The particle mass after the machine had satisfactory friability, the fractions of wood and fibre parts practically did not differ in geometrical parameters.

When transported to the drying chamber, the mass was evenly coated and no burning was observed in the drying chamber due to the correctly set drying regime.

No clogging and enveloping of the mixer blades by bast fibres was observed when gluing the chip mass. Laboratory analysis showed uniform osmolisation of particles, no lump formation was observed. Osmolised chips were fed to the moulding machine by means of a belt conveyor.

The moulding machine operating on the principle of pneumatic separation worked satisfactorily. The carpet was poured in a uniform layer. At a chip moisture content of 11 % there was no clogging and hanging of the mass in the machine units. At the trial increase of moisture content up to 14%   there was observed

clogging of the needle roller and registers. The humidity of the osmolised chips, according to the regulations, was from 8.8 to 11 %.

The chipboard carpet was fed into a hydraulic press. The obtained wood-plastic composite boards at visual observation had smooth even surface from the existing wood ones differed by dark colouring.

Testing of physical and mechanical properties of WPCP was carried out in a specialised laboratory. The test results given in Table 5.3 indicate their compliance with GOST for DSP 10632-77.

Figure 5.3 shows the technological scheme for obtaining wood-plastic composite board materials.

According to the technological scheme, it includes the following types of equipment:

During the production of wood-plastic composite board materials and grinding of initial raw materials a significant amount of aerosol is emitted into the working area.

In order to reduce environmental pollution by dust emissions, we have provided for sealing the roof of the building. When conducting the technological process, it is necessary to strictly observe the safety rules applicable in the industry and comply with the fire safety rules.

During the production of wood-plastic composite board materials it is necessary to observe the rules of industrial sanitation and fire safety according to GOST 12.4.005-88.

Persons associated with the manufacture of wood-plastic composite board materials shall be provided with appropriate personal protective equipment in accordance with GOST 12.4.001-89.

The technological scheme presented in the technological regulations does not have large emissions, technological safety requirements are fully taken into account and protection of the air environment is ensured. Waste recycling is also envisaged for obtaining some types of construction components - material slices.

Production shall be provided with the following instructions:

1. Instruction on safety and industrial sanitation at the workplace.
2. Job description for the plant supervisor.
3. Job description for a shop technologist engineer.
4. Job description for the laboratory technician of the workshop.

We have developed technical specifications TU Uz - 10-90 2018. Wood-plastic composite board (2018).

These specifications apply to wood-plastic composite boards (WPCB), manufactured on the basis of cotton stalk shavings and modified urea-formaldehyde resin. WPCPs can be used in construction industries.

Wood-plastic composite boards (WPCB) should meet the requirements of these technical specifications and be manufactured according to the technological regulations approved in the established order.

In terms of physical and mechanical properties, the boards obtained from cotton stalks should meet the requirements given in Table 5.3.

Table 5.3 **Physical and mechanical properties of slabs obtained from stems**

Physical and mechanical properties	Measurement unit IJA	Wood-plastic composite board made of cotton stalks and binder field		cotton chipboard according to GOST 10632-77
		laboratory	factory	
Density	kg/m^3	620-728	600-750	550-750
Bending strength	MPa	13,6-18,2	10,1-17,6	14,7-17,6
Tensile strength of the plate	MPa	0,5-0,8	0,3- 0,55	0,31-0,35
Swelling	%	16-30	24-34	20-30

Raw materials and materials used for the production of DPKP should comply with the requirements of normative documents approved in accordance with the established procedure and authorised for use by the Ministry of Health of RUz.

WPCPs are packed in special containers designed for transportation to the destination place. Upon agreement with the customer it is allowed to pack WPCPs in other containers. Packaging containers shall be labelled with a paper label stating:

name of the manufacturer or its trade mark, address;

the name of the product and its brand;

batch numbers;

date of manufacture;

shelf life and storage conditions;

net weight;

the designations of the present TSh;

purpose and method of application;

conformity mark (in cases stipulated by the ND NSS RUz);

"O'zbekistonda ishlab chiqarilgan" when sold within the republic, when exported "MADE IN UZBEKISTAN".

Transport labelling according to GOST 9980.4

All work should be carried out with the supply and exhaust ventilation and local suction switched on.

Modified urea-formaldehyde resins are used in the process of production of WPCPs, in case of skin contact, these resins can cause severe burns. Employees should work in special clothes according to GOST 12.4.100 and protective glasses according to GOST 12.4.153. If the resins get on the skin, it should be washed off with 3% boric acid solution.

It is forbidden to handle open flames and other sources of ignition when making the boards in production conditions. Artificial lighting must be of explosion-proof design.

Acceptance rules - according to GOST 9980.1.

The volume of samples - according to GOST 9980.1.

To determine the compliance of the PPCP with the requirements of these technical specifications, acceptance, periodic and certification tests are carried out.

Certification tests are carried out for compliance with all indicators of the present TSh in an accredited laboratory according to ND NSS RUz.

Test methods DPKP is carried out in accordance with GOST 10632-77 according to the following indicators:

bending strength, MPa -17, 0-27,0

tensile strength of 1 pl., MPa - 0.2-1.0

density, kg/m^3 - 650-800

swelling, % -16-30

WPCPs shall be transported by any means of transport in compliance with the transport regulations applicable to that type of transport.

WPCPs are stored in closed warehouses, observing fire safety requirements, at storage temperature from minus -5 to + 50 °C. Guaranteed storage period of DPKP is 12 months from the date of manufacture.

§ 5.4. Calculation of technical and economic efficiency from the application of developed wood-plastic board materials in production conditions

Calculation of economic efficiency of production of wood-plastic composite board materials from cotton stalks and polymer binders was carried out taking into account the conditions of LLC "PROSPER ALL", which is a specialised enterprise for the production of wood-plastic boards, including those made of guza-paya (cotton stalks) and polymer binders. The total volume of product sales is - 20 thousand m^2. We have produced 15 thousand metres^2 pilot batch of wood-plastic composite board materials.

The wood-plastic composite board (WPCB) developed by us costs 20723 UZS per 1 m^2. The thickness of the board ranges from 12 to 18 mm. The generic demand for wood-plastic composite board materials in the mechanical engineering, furniture and construction industries of the republic is several hundred billion square metres.

At present, as noted above, various cookers are still used for this purpose, which are expensive and bought from abroad, for foreign currency.

Cost of Russian slab41344 Sum per 1 m^2.

The calculation of technical and economic efficiency was carried out according to the following formula:

$E^=$ Uobsh (Srp - Sdpkp), Ede Uobsh is the total annual volume;

Srp is the cost of the Russian cooker;

Cdpcp - cost of wood-plastic composite boards.

Consequently, the profit as a result of the use of the developed wood-plastic composite board material from the application of 15000 m^2 in the construction enterprise AZIMUT-MY LLC only due to the price difference, not taking into account the increase in service life, will be: 3=15000(41344- 20723) = 309315000 sum.

And with the production and use, taking into account the needs of the construction and furniture industry in the amount of 100 thousand m^2, the economic effect will be 2062210000 sum. This means the saving of foreign currency funds, in US dollars.

§ 5.5. Conclusions to Chapter Five

The technological line for production of wood-plastic composite board materials using fillers from cotton stalks and urea-formaldehyde resin has been developed and created. Pilot production tests and mastering of this technology were carried out, and the production of a pilot batch of wood-plastic composite board materials was organised.

According to the results of work the technological regulation for manufacturing of composite board material from cotton stalks in industrial conditions and enterprise standard (technical conditions) for them are developed.

The developed technology allowed to realise the waste-free use of cotton stalks in the production of composite wood-plastic board materials and ensured the efficiency of all units and units of the technological line. The results of pilot testing of the obtained board materials showed high physical and mechanical properties that meet modern requirements. The technical and economic calculation on application of this technology has been carried out.

According to the results of the research, the developed "Technological regulations and TU" are implemented at the enterprise LLC "PROSPER ALL" for industrial implementation.

CONCLUSION

1. The scientific and substantiated approach of creation of technology of grinding and obtaining of conditioned wood-fibre mass of fillers from cotton stalks for obtaining of composite wood-plastic board materials on their basis with higher physical and mechanical properties is developed.

2. A two-stage method of obtaining dispersed particles of wood fibre pulp at the second stage of grinding was proposed, at the first stage a two-stage method of obtaining fragments of a certain quality and length (cotton spikes) was used when grinding cotton spikes.

3. Stem moisture content, transfer rate, cutting speed, pile density in primary crushing were determined and in secondary crushing the length and moisture content of cotton stalks were determined, minimum water absorption and absorption coefficient were recommended at 2025 mm and thickness 0.3-0.8 mm.

4. Cutting power up to 100 m/s was revealed, obtaining of conditioned cotton stalks by quality sawing of cotton stalks increased by 30-40%.

5. The maximum value of flexural strength (24 MPa) of composite wood-board materials was found to be observed at an average chip length of 20-30 mm, and the maximum perpendicular strength (0.75 MPa) was observed at an average chip length of 25-30 mm.

6. The relationships between the quality of secondary crushing and physicomechanical properties of composite board materials in the process of obtaining masses of condensate spikes from cotton stalk were revealed.

7. The optimal technological modes of pressing of composite wood-plastic board materials have been recommended, and the standard and technological regulations of the enterprise for their production have been developed.

LIST OF REFERENCES

1. Decree of the President of the Republic of Uzbekistan No. UP-4947 "On the Strategy of actions on five priority directions of development of the Republic of Uzbekistan in 2017-2021".

2. Bataev A.A., Bataev V.A., ed. Composite Materials: Structure, Production, Application, Logos, 2006. 28-29.

3. Karasev E.I. Development of wood-based panels production. / Study guide for universities. M. MGUL. 2001. -C. 3-10; 89-93.

4. Surovtseva L.S. Technology and equipment for production of composite wood materials. //Textbook for universities. Publishing house of Arkhangelsk State Technical University, 2001. -C. 3-7; 180-210.

5. Deppe HJ., Ernst K. MDF - mittitteldichte Faserplatten./ DRW-Verlag. 1996. 200 s; -C. 10-21; 102-120.

6. Deppe HJ., Ernst K. Taschenbuch der Spanplattentechnik (4. Auflage)./ DRW-Verlag, Stutgart, BRD. 2000. -C. 5-10; 112-164.

7. European Panel Federation (EPF). Annual Report. 2001 - 2002. -C. 3-6; 282311.

8. Klyosov A.A. Wood-polymer composites.// Sb. Scientific bases and technologies. 2010. -C.5-14; 210-272.

9. Kholmurodova D.K. Development of effective composition of composite wood-plastic board materials on the basis of stems of annual plants and polymeric binders and technology of their production. Abstract of doctoral thesis (DSc). State Unitary Enterprise "Fan va tarakkiyot", Tashkent - 2021, 50 p.

10. Breitenbach P. Installation for continuous production of bogasse boards,-Bison-Ward Ultra-board Advertising Prospectus / Afkol Group, Maylane, South Africa, 1990. -C. 1-3.

11. Espaeva A.S. Technology of plate materials. Almaty Textbook Pasobia, 2011,-pp.49-59

12. Kholmurodova D.K., Composite wood-plastic boards from cotton stalks as a building material // Composite Materials. - Tashkent, 2015. - №2. -C. 85-86 (02.00.00; №4).

13. Volynskiy V.N. Technology of chip and fibre wood boards. // Study guide for universities. Tallinn. "Desiderata". 2004. -C. 4-6; 112-162.

14. Kholmuradova Dilafruz Kuvatovna, Abed Nodira Soyibjonovna, Negmatov Sayibjan Sadikovich, Mihriddinov Riskiddin, Askarov Kudrat Askarovich. Physical and Mechanical properties of structural composite Wood-Plastic panel Materials depending on the size of particles of wood fractions of fillings from Cotton Stems // European Applied Sciences. - 2016, №9, p.38-40.

15. Boydadaev M.B. Development of effective technology of obtaining composite wood-plastic board materials for construction purposes. Author's abstract of doctor of philosophy (PhD). State Unitary Enterprise "Fan va tarakkiyot", Tashkent - 2019, 42 p.

16. Kholmurodova D.K., Negmatov S.S., Abed S.J., Askarov K.A., Saidov M.M.,

Abdullaev M.B., Buriev N.I. Development of waste-free technology for obtaining composite board materials from cotton stalks // Composite Materials. - Tashkent, 2015. - №3. -C. 79-80 (02.00.00; №4).

17. Siu Yuach-Chjo Research of questions of technology of production of plates from crushed bamboo. Auth. kand. dis,- L.:1962. - 22 c.

18. Leonovich A. A. Technology of wood-based panels: progressive solutions. // Study guide. - SPb. Chemizdat. 2005. -C. 4-6; 182-200.

19. Kholmurodova D.K., Negmatov S.S., Askarov K.A., Mihridinov R.M., On the development, application and organisation of production of composite board materials using stems of annual plants and polymer binders // Composite Materials. - Tashkent, 2016. -№2. -C. 85-86 (02.00.00; №4).

20. Patent No. 2295834.France: B29 5/60Materials substituting wood and method of manufacture.

21. Dilafruz Kholmuradova, Nodira Abed, Komila Negmatova, Kudrat Askarov, Sayibjan Negmatov. Development of Antifrictional and Wearproof Composite Polymeric Materials and Research of Durability of Details of Working Bodies of Cars From Them Working at Interaction With Cotton Raw // Advanced Materials Research, Vol.1145, pp 163-165 doi: 10.4028/ www scientific, net/ AMR.1145.163, 2017. Trans Tech Publications, Switzerland (05.00.00; no.1).

22. Kholmurodova D.K., Negmatov C.C., Mihridinov P.M., Askarov K.A., Abed N.S. Development of optimal technological modes of pressing wood-plastic composite boards from cotton stalks // Composite Materials. - Tashkent, 2017. -№2. -C. 87-88 (02.00.00; №4).

23. Kholmurodova. D.K., Boydadaev M.B., Negmatov. S.S., Abed. N.S. Research and obtaining compositions of composite wood-plastic board materials on the basis of local raw materials and production wastes. Composite materials.-Tashkent, 2018. -№4. C. 97-98 (02.00.00 №4).

24. Trishin SP. Technology of wood-based panels. // Textbook for economic specialities of universities. - M. MGUL. 2001. -C.100-162.

25. Boydadaev M.B., Kholmurodova. D.K., Negmatov. S.S., Abed. N.S. Study of influence of pressing time and pressure, composite wood-plastic board materials on their fracture and water absorption. Composite Materials. -Tashkent, 2019. -№1 C 108-109 (02.00.00 №4).

26. Kholmurodova D.K., Negmatov. S.S., Boydadaev M.V. Esearch influence of humidity of resined screw-polymer weight on parameters of physical and mechanical properties of composite wood and plastic plate materials. International Journal of Advanced Research in Science, Engineering and Technology, Vol.6, Issue 8, August 2019 ISSN:2350-0328. (05.00.00 №8).

27. Glukhikh B.B., Mukhin H.M., Shkuro A.E., Buryndin V.G. Receipt and application of products from wood-polymer composites with thermoplastic polymer matrices:. //Teaching Manual. -Ekaterinburg: Ural. gos. lesotechn. Univ. of Ukraine, 2014. -C. 3-10; 80-84 c.

28. GOST -10632-2007   . Wood chipboards. Technological conditions. M. Standardinform. 2007. -C. 2-5.

29. Negmatov S.S., Boydadaev M.B. Process of manufacturing wood-plastic boards from cotton stalks. //Materiali XV international scientific and practical conference bdedescheto vneshchto voprosi ot lucid na naukata - 2019. 15 - 22 dekemvri 2019 volume 14 Sofia, Bulgaria "Byal GRAD-BG ODD" 2019. -C 127131.

30. Formation of particleboard on the basis of modified phenol-formaldehyde resin. //Dissertation for the degree of Candidate of Technical Sciences. Kostroma, 2016, -C. 11-30, 107-118.

31. Negmatov. S.S., Boydadaev M.V. Z 40 Zbior artykulow naukowych z Konferencji Miedzynarodowej NaukowoPraktycznej (on-line) zorganizowanej dla pracownikow naukowych uczelni,jednostek naukowo-badawczych oraz badawczych z panstw obszaru obszaru bylego Zwiazku Radzieckiego.Warszawa. Poland. *ISBN: 978-836640122-8.*

32. Zavrazhnov A.M. et al. Forest use of wastes, processing of annual plants as raw materials for the manufacture of board materials // Improvement of technology and equipment for the production of wood-based panels. Collected Works "Soyuznauchplit - prom" VNLDrev. Balabanovo, 1983, - P. 54606.

33. Burdin, N.A. et al. Lesopromyshlennyi kompleks: sostoyanie, problemy, perspektivy. - Moscow: Izd-voor MGUL, 2000, -C. 5-12.

34. Volynsky V.N. Technology of chip and fibre wood boards. Textbook for universities Tallinn: Desiderata 2010,- P. 19-20. 19-20.

35. V.N. Volynsky Technology of chip and fibre wood boards Textbook for universities DESIDERATA 2004. -c 28-56.

36. Korpak Y.N. Physical and mechanical properties of annual shoots of grape bushes //Problems of mechanisation and economy of agriculture. - issue 1, - Krasnodar, 1969, -C. 32-38.

37. Popova K.A. Research of pressing process and physical and mechanical properties of boards from debarking waste with binder. // Author's abstract of candidate dissertation. -Л.,1979. - C. 19.

38. Golubitskaya G.A. Some properties of cotton stalks and some physical and mechanical parameters of construction plastics from particles of cotton stalks // Abstracts of the report of the XUL scientific conference of Az PL, - Baku, 1967, - P. S. 22.

39. Special issue on agricnltyral waste utilisation/small industry journal Qivzon City. Philippiness, - 1976. - P. 9-11.

40. Golubitskaya G. A. Study of production technology issues and construction properties of plastic from particles of cotton stalks.// Abstract of Cand. diss. Baku, 1976, -C. 6-9.

41. Rasev A.I., Kosarin A.A., Krasukhina L.P. Technology and equipment of protective wood treatment Textbook. - M.: MGUL, 2010., - PP. 4-12.

42. Kurdyumova V. M. Research and development of technology for the manufacture

of cotton stalk platys. L., 1981, 17 pp.

43. Production of boards from cotton husks Agricultural Wasten an Instematinal Jomd 1985,- vol 13, no. 4, -p. 287-293.

44. Complex utilisation of secondary material resources in Uzbekistan // Report of the scientific party. Conf. 1-2 Apr.1985,- Tashkent, 1965, - P.6770, 107-109, 183.

45. Zhukov. P., Filonov A. A. The use of husk//Mechanical processing of wood,- 1969, № 3, -S.18-20.

46. Filonov A. A. Study of the possibility of replacing wood raw materials in the production of chipboards with sunflower husk, - Author's abstract of candidate's thesis, - Voronezh: 1970, -C. 3-8.

47. Brikmon "Banmrwollsten-Rostoll Sur Vondie Spanuplaten" Bison-werke Bahke and Jreten. 1979. -C 10-14.

48. Otlivanchik A.N. Production and application of particleboards, - Goslitsgroyizdat. M.: 1962. -C.108-152.

49. Amalitsky V.V., Amalitsky V.V. Woodworking machines and tools Textbook. - M.: IRPO, Academy, 2002. -C.12-14.

50. Usmanov H.U., Minina V.S., et al. Prospects of chemical processing of cotton wastes. - Tashkent. Nauka, 2005. -126 c.

51. Negmatov S.S., Atakhodjaev L., Kholmuradova D.K., Lysenko A.M., Abdullaev M.B., Mukhitdinov Z.N. Existing methods and devices for grinding wood and wood-like raw materials to obtain composite wood-plastic materials and their analysis. RITC New composite materials based on organic and inorganic ingredients. 27-28 September 2012. -C. 238-241.

52. Ladan J. Naimi and others Cost and Performance ofWoody Biomass Size Reduction for energy production : CSBE/SCGAB 2006 Annual Conference Edmonton Alberta, July 16-19, 2006. Paper No. 06-107.

53. Banks CJ and others Particle size requirements for effective bioprocessing of biodegradable municipal waste : Technology Research and Innovation Fund Project Report, Defra TRIF Programme, report October 2008 - January 2010. p. 22.

54. Technological instruction for production of particleboards on modified lines SP25 and SP35. -Balabanovo, Kaluga region, VNIIdrev. - 1987. - 101 c.

55. Golubitskaya G.A. Research of questions of production technology and construction properties of plastic from particles of cotton stalks. // Abstract of Cand.dis. Baku, 1976, 18 p.

56. Zavrazhnov A.M. Ways to use wastes in the production of slabs //Express-information: Platyyifanera.M.:   1981.-vp.9
(VNIPIEIlesprom), -C.8-11.

57. Baum M.Yu., Novak A.P. Production of chipboards from grapevine //Fanera and boards, - 1974, - № 10.-S.9-10.

58. A. c. № 125442 USSR MKI B29 5/00. Method of manufacture of straw slabs.

59. A. c. № 656878 USSR MKL B29 5/00. Method of obtaining plates from vegetable raw materials.

60. Application No. 1943969 of the Federal Republic of Germany for cl. B29 5/00. Method and device for obtaining chipboards from stems of annual plants.

61. Ugolev B.N. Woodworking and forest commodity science, Textbook for secondary vocational education. - 2nd ed., erased. - M.: Academy, 2006-S. 9-12, 102108, 212218.

62. Nazirov N. Science and cotton.// Izdvo Uzbekistan,- Tashkent: 1977, 27 p.

63. Negmatov S.S., Kholmurodova D.K., Saidov M.M.. State of use of agricultural wastes and stems of annual plants for the production of composite wood-plastic materials // RMNTC "Nanocomposite materials". -Tashkent, 10-11 April 2009. - C. 112-114.

64. Maksudov N., Kattakhodzhaev R., Tadzhiev E. Technology of waste-free production// Cotton growing. Apromizdat, M., 1985, № 11. - 27c.

65. Mikheev G. Wastes of cotton growing - for fodder for cattle // Cotton growing. - Agropromizdat. 1983, -№7,- C. 12-14.

66. Shoev S. Complex of waste-free production // Cotton production. Agropromizdat. -1983,- № 7, -C. 22-24.

67. A Volynsky V.N. Interrelation and variability of physical and mechanical parameters of wood. Arkhangelsk, ed. AGTU. 2000., - C.4.

68. Volynskiy V.N. Catalogue of woodworking equipment produced in the CIS and Baltic countries. (3rd ed.). M. ASU-Impulse. 2003., -C. 3254.

69. Usmanov H.U., Minina V.S. et al. Prospects of chemical processing of cotton wastes. - Tashkent: Nauka, 1964. -126c.

70. Kasimov N. Forgotten guza-paya //Pravda Vostoka. - 23 November, 1984, p. 3.

71. Kholmurodova D.K. On the possibility of using cotton stalks in various industries. // Nanocomposite materials, Materials of RMNTC of young scientists, Tashkent, -10-11 April 2009. -C. 124125.

72. Kholmurodova D.K., Negmatov S.S., Askarov K.A., Abed S.J., Saidov M.M. Use of agricultural wastes and stems of annual plants in the production of composite wood-plastic board materials and methodological assessment of their properties. - 60c.

73. Kudilov V., Petrov P. All reserves - in business // Cotton growing Agropromizdat, No. 4,- 1985,- P. P. 12-14.

74. Isradilov N., Isradilov T. Cotton stalks are a valuable raw material // Cotton growing. - № 4,- 1986,- C. 9-10.

75. Ermatov I. Utilisation of cotton production wastes with //Cotton growing,- № 4. 1986.-C.10-11.

76. System for generating electricity and steam from cotton husks // Oil mill garetter. 1985, vol. 89. №10. -p.14-16.

77. Lepikh B. Increase the strength of the land // Cotton growing,- № 8,-1983. -C.18-23.

78. Yermatov L.T., Abdullaev A. Waste cotton production in business // Agriculture of Uzbekistan, - 1985. - № 10, - C. 13-14.

79. Nerinsky A.S. Yermatov L.T. On the assessment of cotton production wastes

//Planning and accounting in agricultural enterprises,- 1985. - № 2,- C. 32-35.

80. Paper from guza-pai // Tashkentskaya Pravda. - 4 September, 1981, 2 p.

81. Otlev I.A., Shteinberg C.B. Reference book on particleboards. - M.: Ley, Promst, 1983. - C. 110-142.

82. Gomonai M.V. Resource-saving technologies of wood chopping for wood chips in chopping machines with multi-cutter and knife working bodies: / / / Dissertation of Doctor of Technical Sciences. Khimki. 2003. -C. 90-128.

83. Zalegaller B. G., Lastochkin P. V., Boykov S. P. Technology and equipment of forest warehouses: Textbook for universities - 3rd edition, revised, supplement, - M.: Lesn. pro-mst, 1984. -C. 3-10; 202-262.

84. Puchkov B. V. Grinding of raw materials for wood boards. M.: Lesn. pro- st, 1980. -C. 4-6; 82-88.

85. Modlin B.D., Khatilovich A.A. Production of chips for particleboards. - M.: Lesn. Promst, 1988. -C. 3-9; 101-142.

86. Palgunov P.P., Sumarokov M.V. Utilisation of industrial wastes. - Moscow: Stroyizdat, 1990. - 352 c. (Environmental protection) P. 4-15

87. Tikachev, V. Machines for wood chopping. // LesPromInform. Ser. 68, Techobzor. - 2010. - №2. - C. 92-103.

88. David W. Spencer Large scale rotary shear shredder performance testing, 2010, 10 c.

89. Antimonov C.B., Bulatasov E.O., Ruzavina L.Yu. Obtaining sawdust for mulching in a disc chipper of domestic production: Proceedings of the I All-Russian scientific-practical conference "Innovative technologies in agroindustrial complex: theory and practice", PGSHA, 2013, Penza, -C. 28-30.

90. Sablikov M.N. Choice of type and determination of optimal parameters of the working organ for cotton stalk shredding. // Diss. -Tashkent. 1973, - C. 8-21.

91. Ganiev M.S., Kulametov I.A., Iminjonov B.M. Research of technological process of cotton stalk shredder with drum shredder// Mechanisation of cotton growing. -1978. -№4. -C 1516.

92. Kaplanov A.M. Justification of optimal parameters of working bodies of machines for shredding and processing of cotton stalks. //Dissertation of Doctor of Technical Science -Rostov-on-Don. 1991. -C. 4-11; 102-189.

93. Sablikov N.V. Resistance to cutting of cotton stalks. //Proceedings of TIIIMSH. v. XIX. -Tashkent, 1962, -C. 61-65.

94. Volynskiy V.N. Technology of chip and fibre wood boards. Study guide for universities. Tallinn. "Desiderata". 2010.. -C. 5-12, 201-238.

95. Atakhodjaev A. Justification of the design scheme and parameters of the working bodies of machines for shredding cotton stalks. /Dissertation. - Tashkent, 1987, -C. 3-8.

96. Kaplanov A.M. Analysis and synthesis of working mechanisms of the complex of machines for harvesting and shredding of cotton stalks. /Dies. -Tashkent, 1987, -C. 818.

97. Negmatov S.S., Saidov M.M., Atakhldjaev A. Copyright certificate. №114952. Device for shredding plant stems in bales. SSSR, MKI V27 11/06. 1984г.

98. Negmatov S.S., Kholmurodova D.K., Abed S.J., Buriev N.I., Askarov K.A., Saidov M.M., Atakhodjaev L., Abdullaev M.B. Technology of obtaining fillers from cotton stalks for the production of composite wood-plastic materials. // Tashkent, GUP "Fan va tarakkiyot", 2010, 24 p.

99. Negmatov S.S., Saidov M.M., Tulyaganov B.H., Kholmuradova D.K., Lysenko A.M., Abdullaev M.B., Mukhitdinov Z.N. Technology of obtaining filler from cotton stalks for the production of composite wood-plastic materials. // RSTK New composite materials based on organic and inorganic ingredients. 27-28 September 2012, -C. 30-32.

100. Gribenchikova A.V. Material science in the production of wood boards and plastics. Textbook for technical schools. M. Ley, Promst. 1988, -C. 52-56.

101. Shvartsman G.M. Production of particleboards. - M.: Lesnaya promyshlennost', - 1997. -C. 30-32.

102. Karasev E.I. Equipment of enterprises for the production of wood-based panels. Manual for universities. M. MGUL. 2002.. -C. 48-52.

103. Mezentsev A.V. Plastic from guza-paya //Agriculture of Uzbekistan,- 1972, - № 9, -C. 32-33.

104. Mezentsev A.V., Lazareva AD. Wood boards and plastics // Proceedings of ULTL. - otl. 30. - Sverdlovsk, - 1973, -C. 42-44.

105. Leonovich A. A. Technology of wood-based panels: progressive solutions. Study guide. - SPb. Chemizdat. 2005, -C. 32-34.

106. Dyskin L.M. Dependence of pressing time of chipboards on types of applied wood particles // Woodworking industry,- No. 5,- 1961, - P. 8-9.

107. Razinkov E.M. Production of wood boards and plastics. Textbook for universities. Voronezh. Izd. VGLTA. 1998, -C. 18-19.

108. Shaglev N.A., Shagleva M.F. About some physical and mechanical properties of cotton stalks // Mechanisation of cotton growing. - 1982,- № 8.- C.9

109. Brikman "Baumuwollstengel - Rohstoff fur vondil Spannplaten" Bison - Werke Bahre und Jreten, 1979, p. 28-30.

110. Wood fibreboards based on cotton stems and grape shoots. Cafram Siempelkanp advertising leaflets, 6 p.

111. Kholmurodova D.K. Selection and justification of research objects for obtaining composite wood-plastic materials // RITC Composite materials based on technogenic wastes and local raw materials: composition, properties and application" 15-16 April 2010. -C.28-29.

112. Saidov M.M. Features of the use of cotton stalks in composite press materials // II All-Union Symposium "Biotechnical and chemical methods of environmental protection. Proc. Samarkand, -1988, -C. 84.

113. Kholmurodova D.K., Negmatov S.S., Saidov M.M., Tulyaganov B.H. Factors affecting the formation and magnitude of physical and mechanical properties of

composite wood-plastic materials and boards. // RITC Composite materials based on technogenic wastes and local raw materials: composition, properties and application" 15-16 April 2010. -C. 32-33.

114. Bazhenov V.A. Technology and equipment for production of wood boards and plastics / 2nd edition, revision and supplement. - M. : Ecology, 1992. - 416. -C 4-35.

115. Brutyan K.G. Formation of low-toxic wood materials using adhesives modified with shungite sorbents: Abstract of the thesis of Candidate of Technical Sciences. - SP: 2010. - 20 c.

116. Surovtseva L.S. Technology and equipment for production of composite wood-based materials. Textbook for universities. Publishing house of Arkhangelsk State Technical University, 2001. -21 c.

117. Dyskin, I.M. Cooling and conditioning of particleboards / I.M. Dyskin, I.A. Otlev // Plywood and boards: improvement of particleboard quality. Collection. - Moscow: TsNIITEIlesprom, 1967. - C. 2022.

118. Kononov G.N. Chemical processes occurring during hot pressing in the structure of particleboard based on furfuralacetone monomer FA // Bulletin of Volga Region State Technological University. Series "Forest. Ecology. Nature Management": scientific journal. - Yoshkar-Ola: PSTU, 2013,- № Z.-S. 65-71.

119. Leonovich A.A. Technology of wood boards progressive solutions / A.A. Leonovich. SPb.: KHIMIZDAT, 2005. -C. 4-11, 108-162.

120. Meloni T. Modern production of chipboard and fibreboard / Translated from English by V.V. Amalitsky and E.I. Karasev. Amalitsky and E.I. Karasev. - M. :Lesn. Promst, 1982. -C. 5-15, 142-165.

121. Modlin B.D. Production of shavings for chipboard / - M.: Ley, Promst, 1988. - C. 3-6, 82-97.

122. Otlev, I.A. Intensification of wood chipboard production /M.: Ley, Promst, 1989. -C. 3-15, 82-104.

123. Otlev I.A., Dyskin I.M. et al. Pressing of particleboard at high temperatures. // "Boards and plywood". - VNIIPIEEIlesprom. Express-inform. - M.: Vyp. 5.-12c.

124. Pilzer M.S. Formation of chipboard packages with pneumatic fractionation of wood particles // New in the technique and technology of particleboards: collection of papers. - M.: Lesn. Promst, 1972. - C. 3037.

125. Pozhitok A.I. Theoretical and experimental study of intensification and optimisation of the process of pressing of chipboard: dissertation of candidate of technical sciences - M.: MLTI, 1978. -C. 5-16, 200-209.

126. Sosnin M.I. Physical bases of pressing of particleboard / M.I. Sosnin, M.I. Klimova. - Novosibirsk: Nauka, Siberian Branch, 1981. - C. 3-10,91-102.

127. Tuluzakov D.V. Formation of wood chipboard strength in the process of pressing: dissertation of candidate of technical sciences / - M.: MLTI, 1991. - 366 c.

128. Fedotov A.A. Technology of chipboard boards with enhanced physical and mechanical properties based on furan oligomer: Cand. Sci. (Techn.) Dissertation. - M.: MGUL, 2013. -C. 3-10, 102-154.

129. Chubinsky A.N. Formation of wood particleboards of reduced toxicity // / Izvestiya St.-Peterburgskaya lesotechnicheskoy akademii. - Vyl. 186. SPb.: SPBLTA, 2009. - C. 156-163.

130. Chubinsky A.N. Formation of low-toxic particleboards and using modified adhesives // Forest Journal". - Arkhangelsk: SAFU, No. 6, - P. 67-73.

131. Shvartsman G.M. New tendencies in the production of chipboard. //Obzor. - M.: VNIPIEIlesprom, 1976. -C. 3-8, 20-48.

132. Shvartsman G.M. Production of particleboards. - M.: Lesnaya Promyshlennost', 1977. -C. 4-21, 201-228 c.

133. Shestakova E.Ya. Research of the contacting process of wood particles at gluing of particleboards: Cand. Sci. (Techn.) / E.Ya. Shestakova. - Л. LTA, 1973. - 19 c.

134. Elbert, A.A. Chemical technology of chipboard / A.A..
Elbert. - M.: Lesn. pro-mst, 1984. -C. 3-20, 112-148.

135. Felby C., Hassingboe J., Lund M. Pilot scale production of fibreboards made by laccase oxidised wood fibres: board properties and evidence for cross linking of lignin. // Enzyme and Microbial Techn. 2002. -Vol.31. -P. 736-741.

136. Kirill Chauzov, Galina Varankina. Investigation on gluing Larch Wood by modified, glue. Development and modernisation of production.//Intemational conference on production engineering. Budva, Cma Gora: Bihac University. 2013. P. 737-743.

137. Remonini C., Pizzi A. Foro Compensati Improved waterproofing of UF Plywood adhesives by melamine salts as glue mix hardeners: System performance optimisation/Holzforsch und Holzververt. 1997. Vol. 1. P. 11-15.

138. Varankina G.S., Chubinsky A.N.. Modification of urea formaldehyde resins shungite sorbents/Development and modernisation of production.//Intemational conference on production engineering. Bihac: Bihac University. 2013. P. 14.

139. Varankina G.S., Vysotskiy A.V. Effective low toxic aluminosilicate fillers for phenol formaldehyde adhesives for plywood and particleboard./ Adhesives in woodworking industry// Zvolen.: 1997. P. 114-120.

140. Modlin V.D. Production of particleboards. - M.: Lesnaya Promyshlennaya Promyshlennost: 1983. -C. 50-250.

141. Otlev I.A. et al. Technological calculations in the production of particleboards. - M.: Lesnaya promyshlennost', - 1979, -C. 3-6, 200-210.

142. Otlev I.A., Shteinberg C.B. Reference book on particleboards. - Moscow: Lesnaya Promyshlennaya Industriya. - 1963. -C. 4-8, 14-16.

143. Shvartsman G.M., Shchedro G.A. Production of particleboard. - Moscow: Lesnaya Promyshlennaya Promyshlennost'. -1987. -C. 8-16, 210-230.

144. Kholmuradova D.K., Negmatov S.S., Lysenko A.M., Saidov M.M., Tulyaganov B.H.,. Mukhitdinov Z.N. Methodological grid of experiments in the study of the influence of individual factors on the physical and mechanical properties of composite wood-plastic board materials. 5-7 May 2011. -C. 416-418.

145.   GOST 10633. Wood chipboards. General rules of preparation and carrying out of physical and mechanical tests. - Moscow: Gosstandart of the USSR: Izd vo standards, 1981. 5c.

146.   Negmatov S.S., Madrakhimov A.M., Abed N.S., Negmatova K.S., Boydadaev M.B., Kholmuradova D.K., Jalolov Sh.N. Development of a method of grinding cotton stalks to obtain conditioned wood fibre pulp for the production of wood-plastic boards // Universum, Technical Sciences, November, 2021, № 11, (92) - P. 80-86

147.   GOST 4.207. "Wood chipboards. System of product quality indicators. Nomenclature of indicators". Introduced 1981.01.01.01.   - M.:
Committee for Standardisation and Metrology of the USSR. Izd. of standards. 1981.-4c.

148.   GOST 10634. Particleboards. Methods of determination of physical properties. - Moscow: Committee for Standardisation and Metrology of the USSR: Izd vo Standards, 1991. - 4 c.

149.   GOST 10635. Particleboards. Methods of determination of strength and modulus of elasticity in bending. - Moscow: Gosstandart of the USSR: Izd voe standards, 1989. - 4 c. GOST 10637 - 88. Particleboards. Method of determination of specific resistance to pulling out of nails and screws. - M.: Publishing house of standards, 1987. -4 c.

150.   Madrakhimov A.M., Jalolov Sh.N., Abed N.S., Negmatova K.S., Negmatov S.S., Kholmuradova D.K., Boydadaev M.B. Development of a method of grinding cotton stalks, allowing to obtain a conditioned wood-fibre mass from cotton stalks that meet the requirements of the production of wood-plastic board materials // Composite Materials, - Tashkent, 2021.№2,-C. 299-300

151.   GOST 10636. Particleboards. Methods of determination of tensile strength perpendicular to the plate of the board. - Moscow: USSR State Committee for Product Quality Management and Standards: Izd-wo standards, 1990. - 4c.

152.   Kholmurodova D.K., Negmatov S.S., Lysenko A.M., Saidov M.M., Tulyaganov B.H., Mukhitdinov Z.N. Methodology for determining the physical and mechanical properties of composite ladder-plastic board materials // ISTC New composite materials based on local and secondary raw materials. 5-7 May 2011, -C. 68-70.

153.   Mindrolsky A.K. Technique of statistical calculations. -M.: Nauka. - 1971. -C. 358-359.

154.   Kholmurodova D.K., Negmatov S.S., Lysenko A.M., Saidov M.M., Tulyaganov B.H., Mukhitdinov Z.N. Methods of processing the results of measurements of physical and mechanical properties of composite wood-plastic board materials // ISTC New composite materials based on local and secondary raw materials. 2011 г. -C.29-30.

155.   Belyi V.A., Vrublevskiy V.I., Kupchinov V.I. Wood-polymer structural materials and products. - Minsk, - Science and Technology. - 1960, -C.58.

156.   Sadykov A.S. Cotton plant - a miracle plant. - Moscow: Nauka, 1985. 128c.

157.   Usmanov H.Ch., Minina V.S., Zaripova A.M. Prospects of chemical processing of cotton wastes. Tashkent, FAS, 1964, 126 p.

158.   Meloni T. Modern production of chipboard and fibreboard. Translation from English. V.V. Amalitsky and E.I. Karasev, M., 1982, Lesnaya Promyshlennaya Promyshlennost', 86-87.

159.   Madrakhimov A.M., Jalolov Sh.N., Abed N.S., Negmatova K.S., Negmatov S.S., Kholmuradova D.K., Boydadaev M.B. Study of cutting speed on the quality of the resulting chips from cotton stalks // International Uzbek-Belarusian Scientific and Technical Conference. Composite and metal-polymer materials for various industries and agriculture. Tashkent, 2020 21-22 May - P. 463-464.

160.   Ivanovsky E.G. Cutting of wood. Forest industry. M.-1975. C. 89-92

161.   Belyi V.A., Vrublevskiy V.I., Kupchinov B.I. Wood-polymer structural materials and products. Minsk. 2001. C. 5-8.

162.   Madrakhimov A.M., Valieva G.F., Negmatov S.S., N.S. Abed, D.K., Kholmuradova D.K., Boydadaev M.B. Study of the influence of moisture content of chlobchatnik stems on the quality of the resulting chips // Composite Materials. - Tashkent, 2021. №3, - C. 207-208.

163.   Madrakhimov A.M., Jalolov Sh.N., Abed N.S., Negmatova K.S., Negmatov S.S., Kholmuradova D.K., Boydadaev M.B. Study of the influence of moisture content of cotton stalks on the quality of the resulting chips // Republican Scientific and Technical Conference. Resource- and energy-saving environmentally friendly composite and nanocomposite materials. Tashkent, 2019, - P. 428.

164.   Madrakhimov A.M., Jalolov Sh.N., Abed N.S., Negmatova K.S., Negmatov S.S., Kholmuradova D.K., Boydadaev M.B. Study of the influence of the degree of density of a bundle of stems on the quality of the resulting chips // International Scientific and Technical Conference. Composite materials on the basis of technogenic wastes and local raw materials: composition, properties and application. Tashkent, 2021 16-17 September - P. 229-230.

165.   Madrakhimov A.M., Jalolov Sh.N., Abed N.S., Negmatova K.S., Negmatov S.S., Kholmuradova D.K., Boydadaev M.B. Results of research on the process of the first and second grinding and development of a modular line for grinding cotton stalks to obtain conditioned shredded wood pulp and fillers from them for the production of composite wood-plastic board materials // Composite Materials. - Tashkent, 2021. №3, - C. 233.

166.   Madrakhimov A.M., Abed N.S., Negmatova K.S., Negmatov S.S., Kholmuradova D.K., Boydadaev M.B.. Technique for obtaining samples of woodplastic composite plate materials for determining their physical and mechanical properties and factors affecting them // Harvard Educational and Scientific Review. Great Britain. International Agency for Development of Culture, Educationand Science 0362-8027 Vol.l. Issue 1 Pages 129-133.   10.5281/zenodo.5734293. November 29, 2021.

167.   Madrakhimov A.M., Abed N.S., Negmatova K.S., Negmatov S.S.,

Kholmuradova D.K., Boydadaev M.B.. Investigation of the process of secondary grinding of chips from cotton stalks and production of chip mass fillers consisting of fibrous part of bark, wood part and the smallest part-dust // FEATURES OF THE DEVELOPMENT OF MODERN SCIENCE IN THE PANDEMIC'S ERA I International Scientific and Theoretical Conference. December 3, 2021Berlin, Germany. 2021 75-79 c.

168.   Bogomolov B.D. Chemistry of wood and basics of chemistry. M: Lesnaya Industriya, 1973, P. 10-11.

169.   Sablikov M.N., Ganiev M.S. About some physical and mechanical properties of crushed mass of cotton stalks. Tashkent, Mechanisation of cotton growing, 1971, No. 8, P. 5-6.

170.   Shaglev N.A., Shagleva M.F. About some physical and mechanical properties of crushed mass of cotton stalks. Tashkent, Mechanisation of cotton production, 1982, No. 8, P. 9-12.

171.   Mavlianov N. Research of the process of pressing and harvesting of cotton stalks. Dis.nasoisk., c.t.n. 1965. C. 16-22, 128.